Home Office/The Scottish Office

Fire Precautions Act 1971

GUIDE TO FIRE PRECAUTIONS IN EXISTING PLACES OF WORK THAT REQUIRE A FIRE CERTIFICATE

Factories, Offices, Shops and Railway Premises

London: HMSO

© Crown copyright 1993
Applications for reproduction should be made to HMSO

First published 1993

First edition Crown copyright 1989

ISBN 0 11 341079 4

Contents

		Page
	Introduction	1

PART I
THE APPLICATION OF THE LAW TO EXISTING FACTORIES, OFFICES, SHOPS AND RAILWAY PREMISES

	DEFINITIONS OF LEGAL TERMS USED IN THE GUIDE	6
CHAPTER 1.	FIRE PRECAUTIONS LEGISLATION — ITS APPLICATION TO PREMISES USED AS FACTORIES, OFFICES, SHOPS AND RAILWAY PREMISES	9
CHAPTER 2.	PREMISES FOR WHICH APPLICATION HAS TO BE MADE FOR A FIRE CERTIFICATE	11
CHAPTER 3.	PROCEDURE FOR FIRE CERTIFICATION	13
CHAPTER 4.	EXEMPTION FROM THE REQUIREMENT TO HAVE A FIRE CERTIFICATE	20
CHAPTER 5.	PROHIBITION NOTICES	23
CHAPTER 6.	OFFENCES	24
CHAPTER 7.	RIGHTS OF APPEAL AND GRIEVANCES	28
CHAPTER 8.	EFFECT OF THE FIRE PRECAUTIONS ACT 1971 ON OTHER LEGISLATION	30
CHAPTER 9.	CROWN PREMISES	31
CHAPTER 10.	NEW BUILDINGS, STRUCTURAL ALTERATIONS AND THE EFFECT OF THE BUILDING REGULATIONS	32

CHAPTER 11.	CONSULTATION	34

PART II
A TECHNICAL GUIDE TO FIRE PRECAUTIONS IN FACTORIES, OFFICES, SHOPS AND RAILWAY PREMISES

	DEFINITIONS OF TECHNICAL TERMS USED IN THE GUIDE	36
CHAPTER 12.	FIRE RESISTANCE AND SURFACE FINISHES OF WALLS, CEILINGS AND ESCAPE ROUTES	39
CHAPTER 13.	ASSESSMENT OF FIRE RISK AND ASSOCIATED LIFE RISK	45
CHAPTER 14.	MEANS OF ESCAPE	51
CHAPTER 15.	MEANS FOR GIVING WARNING IN CASE OF FIRE AND FOR DETECTING FIRE	92
CHAPTER 16.	MEANS FOR FIGHTING FIRE	96
CHAPTER 17.	INSTRUCTION AND TRAINING IN FIRE PRECAUTIONS	105

PART III
ADVICE WHICH A FIRE AUTHORITY MAY BE ASKED TO PROVIDE TO OCCUPIERS OR OWNERS OF FACTORY, OFFICE, SHOP OR RAILWAY PREMISES

CHAPTER 18	THE RESPONSIBILITIES OF MANAGEMENT	110
CHAPTER 19.	ASSISTING PEOPLE WITH DISABILITIES	113
CHAPTER 20.	FLOOR COVERINGS, FURNITURE, FURNISHINGS AND SYNTHETIC MATERIALS, ARTIFICIAL AND DRIED FOLIAGE	118

CHAPTER 21.	GOOD HOUSEKEEPING AND THE PREVENTION OF FIRE	120

APPENDICES

APPENDIX A:	REFERENCES TO BRITISH, EUROPEAN AND INTERNATIONAL STANDARDS	126
APPENDIX B:	TECHNICAL STANDARDS FOR UPHOLSTERED FURNITURE, ARTIFICIAL FOLIAGE, TREES, SHRUBS AND FLOWERS	131

TABLES

Table A:	Minimum fire resistance (integrity in minutes)	40
Table B:	Minimum classes for surface spread of flame	42
Table C:	Occupant floor space factors commonly accepted	56
Table D:	Guidelines on distance of travel in metres	58

DIAGRAMS

Diagram 1:	Example of siting of exits to show the 45° rule	54
Diagram 2:	Example of siting of exits to show the angle of divergence within a room in a normal fire risk shop	54
Diagram 3:	Example of distance of travel in a factory with both high and low fire risk areas	60
Diagram 4:	Example of where the escape route consists initially of one direction only from an office of normal fire risk (see paragraphs 14.17 and 14.29)	61
Diagram 5:	Example of access room showing each of the alternative provisions set out in paragraph 14.18(e)(i) to (iii)	62
Diagram 6:	Example of the sub-division of a corridor by fire doors	64
Diagram 7:	Example of shop or office of two floors with no floor area in excess of 90m^2	66

Diagram 8:	Example of shop or office of three floors with no floor area in excess of 90m²	67
Diagram 9:	Example of shop or office of three floors with no floor area in excess of 280m²	68
Diagram 10:	Example of low fire risk factory	69
Diagram 11:	Example of normal fire risk factory	69
Diagram 12:	Example of single stairways separated from floor areas by (a) a protected lobby or (b) a protected corridor	70
Diagram 13:	Example of by-pass route around a stairway	71
Diagram 14:	Example of separate routes from stairway enclosure to separate final exits	72
Diagram 15:	Example of protected route from stairway enclosure to a final exit	72
Diagram 16:	Example of alternative escape routes from stairway enclosures separated by fire-resisting construction	73
Diagram 17:	Example of a stairway separated from the floor area by a protected lobby	75
Diagram 18:	Example of basement separated from the ground floor by 2 x 30 minutes fire doors, one at the foot of the stairway and one at its head	76
Diagram 19:	Example of basement separated from the ground floor by 2 x 30 minutes fire doors, one between any room in the basement and one at the head of the stairway at ground floor	76
Diagram 20:	Example of defined zone for fire-resisting windows and doors	78
Diagram 21:	Example of fire-resisting protection to lift entrance discharging into dead-end corridor	81
Diagram 22:	Example of a window escape	85

INTRODUCTION

About this guide

1. This guide updates and replaces the first edition published in 1989 by HMSO (reference ISBN 0 11 340906 0). It is issued in connection with the operation of the Fire Precautions Act 1971 as it applies to *existing* places of work which are factory, office, shop and railway premises.

2. The purpose of this guide is to set out reasonable standards for means of escape and other fire precautions in *existing* premises which require a fire certificate. It is not a design guide for new buildings or for those which are "materially" altered or subject to a material change of use within the meaning of the Building Regulations. Such buildings in England and Wales are subject to the requirements of the Building Regulations so far as structural alterations relating to fire precautions and means of escape are concerned. In Scotland such buildings are subject to the requirements of the Building Standards (Scotland) Regulations, which may be satisfied through compliance with the Technical Standards associated with these Regulations (see also Chapter 10).

3. Part I of the guide deals with the legislation relating to fire precautions. It explains which premises used as factory, office, shop and railway premises under the 1971 Act require a fire certificate and which do not. Part II contains guidance on the standard of fire precautions which are required in connection with the issue of a fire certificate. Part III makes recommendations in relation to matters on which a fire authority may be asked to provide advice or take into account when issuing a fire certificate.

4. At various points in the text, certain words are in bold print. This means that the technical term is defined and a list of these definitions is at the beginning of Part II. The references in the left-hand margin in Part I are to sections of the Fire Precautions Act 1971. Reference is also made in Part I of the guide to offences under the 1971 Act. Further details relating to penalties and rights of appeal appear in Chapters 6 and 7.

Who the guide is for

5. The technical standards in this guide (Part II) are directed primarily at fire authorities although they will be of importance to other professionals in the field of fire precautions. They aim to set acceptable national standards of fire safety and encourage consistency of enforcement. At the same time the recommendations in the guide allow for flexibility and the exercise of professional judgement and commonsense. Part I gives an explanation of the main aspects of the Fire Precautions Act 71 and, although it should be appreciated that only the courts can provide a legal interpretation of the Act, this part will be of interest to both technically trained officers and occupiers of premises.

6. The guide (and particularly Part III) will also be useful for employers and managers of factory, office, shop and railway premises. The guide does not attempt to address the specific question of what are adequate and suitable fire safety arrangements for people with disabilities as circumstances will differ according to the premises, the numbers employed and the degree of disability. But it is considered that satisfactory means of escape will be achieved if the appropriate provisions of British Standard 5588: Part 8 Code of practice for means of escape for disabled people, are followed. Chapter 19 provides further advice.

British, European and International Standards

7. Where references are made to British Standards in this guide, these are technical standards published by the British Standards Institution. The standards are those in force at the time of publication. Where buildings used as factory, office, shop or railway premises are brought into use after the publication of this guide, the relevant British Standards will be those current at the time the work is undertaken.

8. It should be noted that where a current British Standard is superseded by any European Standard published by the European Committee for Standardisation (CEN), the existing British Standard will either be withdrawn, or the CEN Standard will be published by the British Standards Institution in their BS EN series. In the case of any such European Standard being cited in the Essential Requirements of a Directive, compliance with that Standard becomes mandatory for the purposes of that Directive.

9. International Standards published by the International Organisation for Standardisation (ISO) or the International Electrotechnical Committee (IEC) are normally published in the BS General Series, with reference being made to the ISO or IEC number.

Special fire risks associated with manufacturing processes

10. Enforcement of fire precautions associated with manufacturing processes and materials stored for those processes or in connection with storage activities is the responsibility of the authority responsible for the enforcement of the Health and Safety at Work etc Act 1974. In such premises, because of the nature of the activity, both the fire authority and HSE Inspectors could be involved in enforcement. An occupier/owner can obtain advice on process fire precautions from the local HSE Area Office.

Guide for Managers/Occupiers and Owners

11. A separate non-technical guide is available. Called "Fire Safety at Work", it has been produced by the Home Departments and is available from HMSO (reference ISBN 0 11 340905 2).

Code of Practice for premises exempt from the requirement to have a fire certificate

12. If the premises are exempt, by virtue of the Fire Precautions (Factories, Offices, Shops and Railway Premises) Order 1989, from the requirement to have a fire certificate or have been granted exemption by the fire authority, the occupier (the owner in a building in multiple occupation) of the premises is still required to provide and maintain a certain level of fire precautions. Practical guidance on how to comply with these requirements is given in the Code of Practice for Fire Precautions in Factories, Offices, Shops and Railway Premises not required to have a Fire Certificate (HMSO, reference ISBN 0 11 340904 4).

Fire Safety and Safety of Places of Sport Act 1987: Section 15

13. Section 15 of the 1987 Act came into force on 1 August 1993. This section enables the fire authority to take account of the provision of automatic means for fighting fire before issuing a fire certificate and to specify details of any automatic means for fighting fire in the contents of that certificate. It does this by removing the qualification in sections 5(3)(c) and 6(1)(d) of the Fire Precautions Act 1971 which restricts consideration of the means for fighting fire to such means as are provided "for use in case of fire by persons in the building".

14. The amendment to section 6(1)(d) of the Act also enables the fire authority under section 6(2)(b) to impose requirements in a fire certificate for securing the proper maintenance of any automatic means for fighting fire for life safety purposes with which the relevant building is provided.

Fire Precautions (Places of Work) Regulations

15. The Home Departments propose to make Regulations under section 12 of the Fire Precautions Act 1971 and issue supporting non-statutory guidance to implement the general fire safety provisions of two European Council Directives (89/391/EEC and 89/654/EEC known respectively as the framework directive and the workplace directive) which deal with general fire precautions in places of work.

16. The proposals identify that factories, offices, shops and railway premises which currently require a fire certificate will continue to need one.

Status of the guide

17. This guide has no statutory force but presents a set of national standards which should enable consistency to be achieved by fire authorities. It applies to England and Wales, and to Scotland, but not to other parts of the United Kingdom.

PART I

THE APPLICATION OF THE LAW TO EXISTING FACTORIES, OFFICES, SHOPS AND RAILWAY PREMISES

DEFINITIONS OF LEGAL TERMS USED IN THE GUIDE

Definition of "factory, office, shop and railway premises"

The Fire Precautions (Factories, Offices, Shops and Railway Premises) Order 1989 (SI 1989 No. 76) modifies the definitions of factory, office, shop and railway premises which are taken from the Factories Act 1961 and the Offices, Shops and Railway Premises Act 1963.

The 1989 Order for the first time, brings within the definition of shop premises certain types of covered markets, in particular converted shop premises and buildings which have been adapted to house semi-permanent stalls. This has been achieved in Article 2(3) by narrowing the definition of covered market place so that the only place excepted from the definition of shop premises is where it is held by virtue of a grant from the Crown or of prescription or under statutory authority.

Railway premises are redefined in the 1989 Order so that buildings include structures within the meaning of the 1971 Act. Article 7 of the 1989 Order excludes from its provisions railway premises subject to regulations made under section 12 of the 1971 Act.

In Article 5 of the 1989 Order the term "at work" has been used in place of the term "employed to work" with the effect of widening the categories of those to be counted for purposes of numerical thresholds. Article 5 defines persons at work to include an individual who works under a contract of employment or apprenticeship, one who works for gain or reward other than under such a contract (eg self-employed person), whether or not he employs other persons, and a person receiving training under arrangements made under section 2 of the Employment and Training Act 1973 or section 2 of the Enterprise and New Towns (Scotland) Act 1990. Additionally, in the case of factory premises, a fire certificate is required where explosive or highly flammable materials (other than materials of such a kind and in such a quantity that the fire authority have determined that they do not constitute a serious additional risk to persons in the premises in case of fire) are stored or used in or under the premises.

Factory premises

The expression "factory" means premises constituting, or forming part of, a factory within the meaning of section 175 of the Factories Act 1961 and premises to which sections 123(1) and 124 of that Act (application to electrical stations and institutions) apply.

The 1971 Act defines premises in section 43(1) as a "building or part of a building" and enforcement is therefore related to factories falling within this definition.

Office premises

Section 1 of the Offices, Shops and Railway Premises Act 1963 defines the premises to which the Act applies and "office premises" means a building (or part of a building) the sole or principal use of which is as an office or for "office purposes". "office purposes" includes administration, clerical work, handling money, and telephone and telegraph operating; and "clerical work" includes writing, book-keeping, sorting papers, filing, typing, duplicating, machine calculating, drawing and the editorial preparation of matter for publication. This definition includes offices which are part of a building used for other purposes (eg offices in factories, hospitals and clubs). It also includes such places as ticket offices, travel agencies and betting offices.

Shop premises

"Shop premises" is defined in terms which include the following:-

(a) Shop (retail)

This includes "shops" in the everyday sense of the word eg butchers, grocers, department stores, etc; it also includes a building (or part of a building):-

(i) of which the sole or principal use is the carrying on there of retail trade or business, eg hairdressers, retail sales by auction, lending books or periodicals for gain; and

(ii) to which members of the public are invited to resort either to leave goods for repair or treatment or to carry out the repair or treatment themselves, eg the parts of shoe repair shops or dry cleaning establishments where goods are received, or self-service laundries.

(b) Wholesale premises or warehouses

This includes a building (or part of a building) occupied by a wholesale dealer or merchant where goods are kept for sale wholesale (but excluding warehouses forming part of factories or belonging to the owners, trustees or conservators of docks, wharves or quays).

(c) Catering establishments open to the public

This includes such places as teashops, cafes, restaurants, public houses and fish and chip shops.

(d) Fuel storage premises

This includes premises used for the storage of coal, coke and similar solid fuel and occupied for the purpose of a trade or business which consists of, or includes, the sale of such fuel.

Railway premises

"Railway premises" means a building (or part of a building) occupied by railway undertakers for the purposes of the railway undertaking carried on by them and situate in the immediate vicinity of the permanent way. This would include, for example, signal boxes and buildings in stations and goods yards where people are employed and structures such as railway platforms. Offices, shops, living accommodation for people employed on the railways, hotels and electrical stations to which section 123(1) of the Factories Act 1961 applies are however excluded.

Chapter 1:

FIRE PRECAUTIONS LEGISLATION – ITS APPLICATION TO PREMISES USED AS FACTORIES, OFFICES, SHOPS AND RAILWAY PREMISES

1.1 The Fire Precautions Act 1971 is the principal instrument for the control of fire safety in occupied premises, and is designed to ensure the provision of adequate general fire precautions, means of escape and related fire precautions in premises within its scope. Fire authorities are responsible for the issue of fire certificates and they have a duty to enforce the provisions of the Act, and the regulations made under it, within their areas. In respect of Crown premises the responsibility rests with the Fire Service Inspectorates of the Home Office and The Scottish Office.

Legislation

1.2 The Fire Precautions Act 1971 came into force in 1972, when hotels and boarding houses were the first class of premises to be designated. Since then all but the smallest of these are required to have a fire certificate. (A separate guide exists for hotels and boarding houses).

1.3 In 1977 the Fire Precautions (Factories, Offices, Shops and Railway Premises) Order 1976 (SI 1976 No. 2009) came into force and required certain of these premises to have a fire certificate. In the case of those factories, offices, shops and railway premises not requiring a fire certificate certain fire safety requirements were imposed under the Fire Precautions (Non-Certificated Factory, Office, Shop and Railway Premises) Regulations 1976. The 1976 Order and Regulations have been revoked by the Fire Precautions (Factories, Offices, Shops and Railway Premises) Order 1989 (SI 1989 No. 76) and the Fire Precautions (Non-Certificated Factory, Office, Shop and Railway Premises) (Revocation) Regulations 1989 (SI 1989 No. 78).

1.4 The Fire Safety and Safety of Places of Sport Act 1987 amended the 1971 Act and provides, among other things, for certain premises to be exempted by a fire authority from the requirement to have a fire certificate. It also gives the fire authority power, in appropriate cases, to serve improvement notices on the occupier or owner of premises. In addition, where there is serious risk to life, the fire authority have the power to issue a prohibition notice without the need to obtain a court order. This is explained further in Chapters 4 and 5 of this guide. Additionally, immediately an application for a fire certificate has been submitted, the owner or occupier as the case may be, has an interim duty to take certain minimum fire precautions. These are outlined in paragraph 3.3.

1.5 Section 3 of the 1987 Act inserted a new section 8B into the 1971 Act which makes provision for the first time for charges to be made by fire authorities to recover the cost of issuing and amending fire certificates.

Health and Safety at Work etc Act 1974

1.6 The Health and Safety at Work etc. Act 1974 (HSW Act) is concerned with securing the health, safety and welfare of persons at work, and with protecting people who are not at work from risks to their health and safety arising from work activities. In terms of fire precautions, the HSW Act and its relevant statutory provisions are used to control the keeping and use of explosive or highly flammable substances and are concerned with precautions against the outbreak of fire in all work activities. Matters falling within the scope of the HSW Act include the storage of highly flammable materials, the control of flammable vapours and dusts, safe systems of work and the control of sources of ignition. In the case of certain types of premises where the processes are of such a nature or on such a scale as to have a direct bearing on general fire precautions, it has been thought right that fire certification of these premises should be dealt with under the 1974 Act. These premises are specified under the Fire Certificates (Special Premises) Regulations 1976 (SI 1976 No. 2003), enforced by the Health and Safety Executive, and include such premises as nuclear installations, explosives factories, any building on the surface of a mine and large petroleum installations. Premises subject to these Regulations are excluded from control under the 1971 Act. Guidance on these Regulations can be obtained from the area office of the Health and Safety Executive.

Premises in multiple occupation

1.7 Schedule 2 to the Fire Precautions Act 1971 was inserted by the 1987 Act and modifies the application of the Act in the case of factory, office, shop and railway premises which require a fire certificate and which either:-

- (a) are held under a lease or an agreement for a lease or under a licence and consist of part of a building all parts of which are in the same ownership; or

- (b) consist of part of a building in which different parts are owned by different persons.

It does this by substituting for references to the occupier of the premises a reference in (a) to the owner of the building and in (b) to the persons who between them own the building. The effect of these modifications in such premises is to enable the fire authority to deal with the owner or owners, as the case may be, in respect of structural and other requirements relating to the building leading to the issue of the fire certificate. It also makes the owner(s) responsible if the premises are used without a fire certificate and for notifying the fire authority of proposed material alterations to premises.

Chapter 2: PREMISES FOR WHICH APPLICATION HAS TO BE MADE FOR A FIRE CERTIFICATE

2.1 By virtue of the Fire Precautions (Factories, Offices, Shops and Railway Premises) Order 1989 the following uses of premises are designated for the purposes of section 1 of the 1971 Act (which requires fire certificates for premises put to designated uses), that is to say:-

(a) use as factory premises;

(b) use as office premises;

(c) use as shop premises; and

(d) use as railway premises,

being (in each case) a use of premises in which persons are employed to work.

2.2 A fire certificate is not by virtue of section 1 of the 1971 Act required for any factory premises, office premises, shop premises or railway premises in which:-

(a) not more than twenty persons are at work at any one time; and

(b) not more than ten persons are at work at any one time elsewhere than on the ground floor of the building constituting or comprising the premises, unless one or more of the conditions specified in paragraph 2.3 applies to the premises.

2.3 The conditions referred to in paragraph 2.2 are:-

(a) that the premises are in a building containing two or more sets of premises which are put to any of the uses described in paragraph 2.1 above and the aggregate of the persons at work at any one time in both or (as the case may be) all those sets of premises exceeds twenty;

(b) that the premises are in a building containing two or more sets of premises which are put to any of such uses and in both or (as the case may be) all those sets of premises the aggregate of the persons at work at any one time elsewhere than on the ground floor of the building exceeds ten; and

(c) that, in the case of factory premises, explosive or highly flammable materials (other than materials of such a kind and in such a quantity that the fire authority have determined that they do not constitute a serious additional risk to persons in the premises in case of fire) are stored or used in or under the premises.

2.4 Any reference to persons at work is a reference to any of the following persons:-

(a) an individual who works under a contract of employment or apprenticeship;

(b) an individual who works for gain or reward otherwise than under a contract of employment or apprenticeship, whether or not he employs other persons; and

(c) a person receiving training provided pursuant to arrangements made (whether before or after the coming into force of section 25 of the Employment Act 1988) under section 2 of the Employment and Training Act 1973 or section 2 of the Enterprise and New Towns (Scotland) Act 1990.

Chapter 3: PROCEDURE FOR FIRE CERTIFICATION

Application for a fire certificate and statutory interim duty

3.1 The Fire Precautions (Application for Certificate) Regulations 1989 (SI 1989 No. 77) prescribe the form of application for a fire certificate (FP1 Revised 1993).

Section 5(1)

3.2 An application for a fire certificate must be made to the fire authority for the area in which the premises are situated, using form FP 1 Revised 1993 with the revised notes issued in 1993. Copies of this form are obtainable from the fire authority.

3.3 Whilst the application for a fire certificate is pending the owner/occupier is under a statutory duty to secure:-

Section 5(2A)

(a) that the means of escape in case of fire at the premises can safely and effectively be used at all times when there are people in the premises;

(b) that the means for fighting fire with which the premises are provided are maintained in efficient working order; and

(c) that any persons employed to work in the premises receive instruction or training in what to do in case of fire.

Consideration of exemption

Section 5(3)

3.4 Where an application has been made for a fire certificate, it shall be the duty of the fire authority to consider whether the premises qualify for exemption under section 5A of the Act and whether or not to grant exemption from the requirement to have a fire certificate. However, an exemption cannot be granted unless an inspection has taken place within the preceding 12 months (see Chapter 4).

Requests for further information

Section 5(2)

3.5 Before issuing a fire certificate the fire authority may need to ask the occupier or, in the case of premises in multiple occupation, the owner for more information. Plans may be required of the premises and if the premises are only part of a building, plans of other specified parts of the building, so far as this is possible. If plans are not provided within the specified time or such further time as the fire authority may allow, the application will be deemed to have been withdrawn. Plans should comprise simple outline drawings showing the essential features although a fire authority will normally accept architects' plans if these are readily available and suitable for the purpose.

Inspection of the premises

Section 5(3)

3.6 Following the application for a fire certificate the premises must be inspected by the fire authority. If the premises form only part of a building, the other parts may also require inspection.

Sections 19, 20 & 40

3.7 An inspector may take such action as is necessary for giving effect to the Act or regulations made under it. This includes power to enter at any reasonable time premises to which the Act applies, as well as the rest of the building containing such premises. Inspectors may also make such inquiries as may be necessary to find out if the Act and any regulations under the Act are being complied with, and may require appropriate facilities and assistance to be given to them in the exercise of their powers.

Section 19(4)

3.8 Inspectors may be required to produce proof of their authority upon request.

3.9 Inspectors should also have evidence of their identity and although there is no national standard identity document for fire brigade officers the document should state the name of the bearer, specify the legislation under which they are empowered to act, and be signed either by the Chief Fire Officer (in Scotland, the Firemaster) or the Chief Executive of the County or Regional Council.

Disclosure of information

Section 21

3.10 The Fire Precautions Act 1971 prohibits inspectors from disclosing any information they have obtained whilst in any premises entered by them in the course of carrying out their duties under the Act, unless such disclosure is necessary in the performance of their duty; for the purposes of legal proceedings or a report of such proceedings; or to an enforcing authority (within the meaning of the Health and Safety at Work etc. Act 1974) to enable that authority to discharge any function falling within its field of responsibility. Details of the issue of prohibition notices are made available for public scrutiny under the provisions of the Environment and Safety Information Act 1988 (see Home Office Circular No. 26/1989 – Fire Service Circular No. 4/1989 (in Scotland Fire Service Circular 3/1989)).

Issue of a fire certificate

Section 5(3)

3.11 If, following the inspection, the fire authority are satisfied that the means of escape from fire and related fire precautions in the premises concerned are such as may reasonably be required in the circumstances of the case, they must issue a fire certificate (but see paragraph 3.4).

Contents of a fire certificate

Section 6(1)

3.12 The fire certificate will specify:-

(a) the use(s) of the premises covered by the fire certificate;

(b) the means of escape in case of fire;

(c) the means for ensuring that the means of escape can be safely and effectively used at all material times (this would cover such matters as measures to restrict the spread of fire, smoke and fumes, escape lighting and directional signs);

(d) the type, number and location of the fire-fighting equipment for use by persons in the building (see paragraph 13 of the Introduction);

(e) the type, number and location of the fire alarms;
and,
in the case of factories;

(f) particulars of any explosive or highly flammable materials which may be stored or used in the premises.

The certificate may include a plan showing any of the above.

Section 6(2)

3.13 The fire authority may also decide to incorporate any of the following requirements in the fire certificate:-

(a) that the means of escape are properly maintained and kept free from obstruction;

(b) that everything covered in (c), (d) and (e) in paragraph 3.12 is properly maintained;

(c) that all employees are given appropriate training in what to do in case of fire, and that records are kept of that training;

(d) that the number of people who may be in the premises at any one time do not exceed a specified limit; and

(e) other precautions to be observed in relation to the risk from fire to persons in the premises.

Section 6(3)

3.14 The certificate may apply any of the above requirements in varying degree to different parts of the premises.

Necessary improvements before a fire certificate is issued

Section 5(4)

3.15 If, on inspection, the fire authority are not satisfied about the matters at (b) to (f) in paragraph 3.12 they must serve a notice on the applicant stating what steps will have to be taken before they are so satisfied, and that they will not issue a fire certificate unless those steps are taken within a specified time. If a fire certificate is not issued by then or within such further time allowed by the fire authority (or by the court in the event of an appeal), the fire certificate will be deemed to have been refused, and it will be unlawful to use the premises for the purpose in question.

Section 9(1)(a)

Section 28

3.16 In general, the applicant is responsible for doing or having done whatever work is necessary but anyone who is prevented by the terms of a lease or agreement from doing what is necessary may apply to a county court (the sheriff in Scotland) for an appropriate order. Provision is also made in such cases for an application to be made to a county court (the sheriff in Scotland) for an order that the expense in whole or part be met by some other person having an interest in the premises and for the modification of the terms of any agreement or lease relating to the premises.

Obligations under fire certificates

Section 6(4)

Section 6(5)

Section 6(6)

3.17 Matters specified in a fire certificate (see paragraph 3.12) must be kept in accordance with that specification, and other requirements of a fire certificate must always be observed. The occupier of the premises covered by a fire certificate will normally be the person responsible in the event of contravention of the requirements of the fire certificate. In the case of premises forming part of a building in multiple occupation the owner of the premises would normally be responsible, but the fire authority may impose requirements on the occupier of the premises in respect of matters over which the occupier, as distinct from the owner, has control.

Section 6(5)

Section 6(6)

3.18 The fire authority also have discretion to impose requirements on other persons instead of, or in addition to, the occupier/owner. The fire authority can also impose requirements relating to a part of the building other than that part comprising the premises and these requirements would be imposed on the occupier(s) of the other part. In these cases the fire authority must consult the other occupier(s) before imposing a requirement on them in respect of a fire certificate. There are rights of appeal regarding the requirements of a fire certificate (see Chapter 7).

Where the fire certificate should be kept

Section 6(8)

3.19 The fire certificate, or in the case of premises forming part of a building in multiple occupation a copy of the certificate, must be kept in the premises to which it relates. It is an offence not to keep the fire certificate in the premises (see Chapter 6).

Charges for issuing and amending fire certificates

Section 8B

3.20 Under the 1971 Act provision is made for fire authorities to charge a reasonable fee for:-

(a) issuing a fire certificate;

(b) amending a fire certificate; or

(c) issuing a new fire certificate as an alternative to amending an existing one.

3.21 This fee, which is determined by the fire authority, covers the cost of work reasonably done but does not include the cost of any inspections carried out.

3.22 There is no charge for amending a fire certificate if the sole cause of the amendment is the coming into force of any regulations made under section 12 of the Fire Precautions Act 1971.

Existing fire certificates issued under the Factories Act 1961 or the Offices, Shops and Railway Premises Act 1963

3.23 By virtue of Schedule 8 to the Health and Safety at Work etc. Act 1974, an existing fire certificate issued under the Factories Act 1961 or the Offices, Shops and Railway Premises Act 1963 is deemed to be a fire certificate under the Fire Precautions Act 1971 so long as there is no material change. The existing certificate shall be treated as imposing, in relation to the premises, the like requirements as were previously imposed by the appropriate fire safety sections of the 1961 and 1963 Acts.

Proposal to change the use of premises

3.24 If there is a current fire certificate for the premises but it is proposed to change the use the fire authority should be consulted before this change takes place.

Changes of conditions and alterations to premises having a fire certificate

Section 8(2)

3.25 The fire authority must be informed in advance if it is proposed:-

(a) to make a material extension of, or material structural alteration to, the premises; or

(b) to make a material alteration in the internal arrangement of the premises or in the furniture or equipment with which the premises are provided; or

(c) in the case of factory premises to begin to store or use explosive or highly flammable materials or materially to increase the extent of such storage or use.

It is an offence to take such action without having given notice of the proposal to the fire authority (see Chapter 6).

Note: See Chapters 10 and 11 in relation to consultation with the Building Control Authority.

3.26 The expression "material" is not defined in the 1971 Act but it is considered that an alteration is material if it would render the means of escape or related fire precautions inadequate in relation to the normal conditions of the use of the premises and the fire risk existing at the time the fire certificate was issued. It is unlikely, therefore, that the fire authority will need to be informed each time it is proposed to redecorate, but there is an obligation to do so if the proposals involve any alterations to the means of escape and its associated matters including decorative changes which adversely affect the surface spread of flame requirements. In case of doubt the fire authority should be consulted.

Sections 8(2) & 8(3)

3.27 The obligation to notify the fire authority of proposals described in paragraphs 3.25 to 3.26 is placed on the occupier of the premises, except in the case of premises forming part of a building in multiple occupation when the obligation is placed on the owner of the building. However, in such cases there is a complementary requirement on the occupier to inform the owner of such proposals they might wish to make.

Section 8(4)

3.28 If the fire authority take the view that the means of escape or related fire precautions will become inadequate if the proposals are carried out they may serve a notice on the occupier (or in the case of premises forming part of a building in multiple occupation, the owner of the building) informing him of the steps that would have to be taken to prevent the matters in question from becoming inadequate. The fire authority may also give such directions as to ensure that the means of escape and associated fire precautions remain adequate whilst the works are being carried out in the premises.

Section 8(5)

3.29 If the fire authority take the view at any time that the means of escape or related fire precautions have become inadequate they may serve a notice on the occupier/owner informing him of the steps that would have to be taken to make the matters in question adequate.

Sections 8(4) & 8(5)

3.30 If in the cases described in paragraph 3.28 or 3.29 the steps are duly taken the fire authority shall, if necessary, amend the fire certificate or issue a new one.

Section 8(5)

3.31 If the steps necessary to restore the means of escape or related fire precautions to a state of adequacy are not taken within a specified time the fire certificate may be cancelled.

Section 8(6) **3.32** The fire authority may also amend the fire certificate or issue a new one if they consider that, in consequence of a change of conditions or of the coming into force of any regulations made under section 12 of the Act, any of the requirements of the certificate need to be varied, revoked or added to; or if the effect of the certificate needs to be altered as to the person(s) responsible.

Chapter 4: EXEMPTION FROM THE REQUIREMENT TO HAVE A FIRE CERTIFICATE

Section 5A (1)
Section 5A (2)
Section 5A (3)
Section 5A (4)

4.1 The Fire Safety and Safety of Places of Sport Act 1987 amended the 1971 Act to enable a fire authority to exempt certain premises from the requirement to have a fire certificate covering a particular use. They may do so following the initial inspection of the premises or at any time after a fire certificate has been issued. In coming to a decision they must take into account all the circumstances of the case and in particular the degree of seriousness of risk in case of fire to the people who may at any time be in the premises. A model exemption notice is attached as Annex B to Fire Precautions Act 1971 Circular No. 16 (in Scotland Memorandum No. 16).

4.2 In considering whether premises may be exempted, the fire authority must work within the provisions of the Fire Precautions (Factories, Offices, Shops and Railway Premises) Order 1989 which specifies the descriptions of premises which qualify for exemption. In the case of a use of premises as factory premises, office premises or railway premises, the premises qualifying for exemption are any premises consisting of or comprised in:-

(a) the ground floor of a building; or

(b) the ground floor and basement of a building; or

(c) the ground floor and first floor of a building; or

(d) the ground floor, first floor and basement of a building in which the basement is separated from the ground floor by fire-resisting construction.

4.3 In the case of a use of premises as shop premises the premises qualifying for exemption are any premises consisting of or comprised in:-

(a) the ground floor of a building; or

(b) the ground floor and basement of a building in which the basement is separated from the ground floor by fire-resisting construction.

For the purposes of the above, construction shall be treated as fire-resisting if, and only if, it is of such a nature as to be capable of providing resistance to fire for a period of not less than thirty minutes.

Section 5A (5)

4.4 The fire authority shall not grant an exemption without an inspection of the premises being carried out unless there has been an inspection within the preceding 12 months.

Section 5A (2)
Section 5A (7)

Section 5A (8)

4.5 An exemption may be granted when an application for a fire certificate is made or at any time when a fire certificate is in force. The fire authority are required to notify the applicant for the fire certificate or the occupier of the premises, as the case may be, that they have granted an exemption. The notice of exemption may include a statement specifying the maximum number of persons who, in the fire authority's opinion, can safely be in the premises at any one time. If it is proposed to make use of the premises in such a way as to exceed this maximum number of persons while the exemption is in force, this is one of the changes of conditions of which the fire authority must be given prior notice under section 8A(1) and (2).

Withdrawal of exemption

Section 5B(1)

Section 5B(2)

4.6 The fire authority who have granted an exemption from the requirement to have a fire certificate covering any particular use of the premises may withdraw the exemption at any time. In reaching their decision, the fire authority must take into account all the circumstances of the case and in particular the degree of seriousness of risk in case of fire to the people who may at any time be in the premises.

Section 5B(3)

4.7 Before withdrawing an exemption the fire authority are required to notify the occupier that they propose to do so, to give their reasons for so doing, and to give the occupier an opportunity of making representations to them on the matter. It is suggested that a period of 21 days is allowed for this purpose. A model notice of proposal to withdraw exemption is attached as Annex C to Fire Precautions Act 1971 Circular No. 16 (in Scotland Memorandum No. 16).

Section 5B(4)

4.8 If the fire authority proceed to withdraw the exemption, they must serve a notice to that effect, after which the occupier has a period of not less than 14 days from the date of service of the notice before the exemption ceases to have effect. A model notice of withdrawal of exemption is attached as Annex D to Fire Precautions Act 1971 Circular No. 16 (in Scotland Memorandum No. 16).

Note: There is no right of appeal to a court against a decision by a fire authority to withdraw an exemption.

4.9 When the exemption is withdrawn a new application for a fire certificate must be submitted if the premises are to continue to be put to a designated use.

Improvement notices

Section 9D (1)

4.10 If, in the fire authority's opinion, any premises exempt from the requirement to have a fire certificate (either by virtue of the designation order or by an exemption granted by a fire authority) are not provided with reasonable means of escape and means for fighting fire the authority may serve an improvement notice on the occupier of the premises.

4.11 In the notice the fire authority should specify (by reference to the Code of Practice if they think fit) what steps they consider are necessary to remedy the contravention. The notice should also specify the period within which such steps should be completed. This must be at least 21 days, which allows the occupier time in which to appeal (see Chapter 7).

Section 9D (2)

4.12 The fire authority may withdraw the improvement notice at any time, normally when they have inspected the premises and are satisfied that the matters have been remedied.

4.13 The fire authority may also grant reasonable extensions of time but they cannot grant an extension if there is an appeal pending regarding the improvement notice as the effect of bringing an appeal under section 9E is to suspend the operation of the improvement notice until the appeal is disposed of or withdrawn.

4.14 The fire authority should obtain proof that the person on whom any of the notices described in paragraphs 4.1, 4.5, 4.6 and 4.9 are served has received the notice.

Chapter 5: PROHIBITION NOTICES

Section 10(2) — **5.1** If, in the opinion of the fire authority, the risk to persons on the premises in case of fire is or will be so serious that the use of the premises ought to be prohibited or restricted, the fire authority may serve a prohibition notice on the occupier.

Section 10(3) — **5.2** The risk may include anything which affects escape from the premises.

Section 10(4) — **5.3** The prohibition notice should only specify those matters which give (or will give) rise to the dangerous conditions. It should direct that the use of the premises to which the notice relates be prohibited, or be restricted to a specified extent, until the matters which the notice has specified have been remedied.

Section 10(5) — **5.4** The notice may include directions as to the steps which will have to be taken in order to remedy those matters.

Section 10(6) — **5.5** The notice will take effect immediately it is served if the risk of serious personal injury is, or will be, imminent. Otherwise the notice will state a period at the end of which it will take effect.

5.6 In serving a prohibition notice, the fire authority may use the standard form attached as an annex to Fire Precautions Act 1971 Circular No. 14 (in Scotland Memorandum No. 14). The fire authority should obtain proof that the person on whom the notice is served has received the notice.

Section 10(7) — **5.7** The fire authority may withdraw a prohibition notice at any time, normally when they have inspected the premises and are satisfied that steps have been taken to remedy the matters specified in the notice. A prohibition notice may also be cancelled as a result of the court's direction following an appeal under section 10A. The bringing of an appeal against a prohibition notice does not suspend the operation of the notice unless the court so directs. However it does suspend any directions issued by the fire authority as to the steps which need to be taken to remove the risk.

Chapter 6: OFFENCES

6.1 This chapter provides a list of offences under the Fire Precautions Act 1971, with the maximum penalty on summary conviction being indicated in paragraph 6.7 by A, B, C or D.

A is level 3 on the standard scale;
B is level 5 on the standard scale;
C is the "statutory maximum"; and
D is the "prescribed sum".

The standard scale

6.2 The Criminal Justice Act 1982, besides amending the Fire Precautions Act 1971, established five levels of fines, rising from level 1 (minimum) to level 5 (maximum), with provision for the value of each level to be increased from time to time.

The statutory maximum

6.3 The "statutory maximum" is a term which derives from section 74 of the Criminal Justice Act 1982. In England and Wales it means the prescribed sum within the meaning of section 32 of the Magistrates' Courts Act 1980. In Scotland it means the prescribed sum within the meaning of section 289(G) of the Criminal Procedure (Scotland) Act 1975.

(The "statutory maximum" always means the prescribed sum).

The prescribed sum

6.4 The "prescribed sum" has a particular meaning in each part of the United Kingdom, by reference to the law of the relevant country. In England and Wales the "prescribed sum" as defined in section 32(a) of the Magistrates' Courts Act 1980 is such sum as is for the time being substituted in that definition by an order in force under section 143(1) of that Act. In Scotland the sum is laid down by an order in force under section 289B(1) of the Criminal Procedure (Scotland) Act 1975.

6.5 All offences marked C and D in paragraph 6.7 are offences which may go to a higher court than a magistrates' court (in England and Wales) or the sheriff (in Scotland), being triable either summarily or on indictment. If the case does go to a higher court and the defendant is convicted on indictment, he or she is liable to a fine, or imprisonment for a period not exceeding two years, or both.

6.6 Under the Fire Precautions Act 1971, except in certain specified cases concerning persons other than the owner/occupier themselves, ignorance of a requirement is not a defence.

6.7 Offences are as follows:-

Sections 7(1) & 7(3)	(a)	putting premises to a designated use when a fire certificate has not been issued and an application for a fire certificate has not been made (D);
Section 7(3A)	(b)	contravening a requirement of the statutory interim duty pending disposal of an application for a fire certificate (B);
Section 7(4)	(c)	contravening a requirement of a fire certificate. The offence is committed by every person who is named in the certificate as being responsible in the event of contravention of that requirement (D);
Section 7(6)	(d)	failing to keep the fire certificate (in premises in a building in multiple occupation, a copy of the certificate) in the premises to which it relates (A);
Sections 8(2)(a) & (b)	(e)	while a fire certificate is in force, proceeding with alteration or change of condition which adversely affects the means of escape or other fire precautions required by the certificate without first notifying the fire authority of the proposal to do so (D);
Sections 8(2)(c) & 8(3)	(f)	while a fire certificate is in force, beginning to keep explosive or highly flammable materials at the premises in greater than the prescribed quantity without first notifying the fire authority of the proposal to do so. (No quantities have, as yet, been prescribed. However, in the case of factory premises the requirement to notify the fire authority relates to proposals to store or use explosive or highly flammable materials in the premises or materially to increase the extent of such storage or use.) (D);
Section 8(7)	(g)	after notifying the fire authority of the proposals, contravening a direction by the fire authority (which they must serve within two months of receipt of the notification) not to carry out a specified proposal until certain steps have been taken (D);
Sections 8A(1) & 8(3)	(h)	in the case of factory premises, while an exemption from the requirement to have a fire certificate is current, beginning to keep explosive or highly flammable materials at the premises without first notifying the fire authority of the intention to do so (C);
Sections 8A(1) & 8A(2)	(i)	while an exemption from the requirement to have a fire certificate is current, proceeding with any alteration or change of condition which adversely affects the means of escape or

		other fire precautions without first notifying the fire authority of the intention to do so (C);
Section 9A(3) & (4)	(j)	while an exemption from the requirement to have a fire certificate is current, contravening the duty to provide such means of escape in case of fire and such means for fighting fire as may reasonably be required in the circumstances of the case (unless the duty is the subject of an improvement notice which has been served) (B);
Section 9A (4)		
Section 9F	(k)	contravening a requirement imposed by an improvement notice (C);
Section 10B(1)	(l)	contravening a prohibition or restriction imposed by a prohibition notice (C);
Section 10B(2)		*Note: It is a defence for a person, other than the one on whom the notice was served, to prove that he or she did not know and had no reason to believe that the prohibition notice had been served.*
Section 12(4)(c)	(m)	contravening any specified provision of any regulations made under section 12 of the 1971 Act (D);
Section 19(6)	(n)	intentionally obstructing an inspector in the exercise or performance of his or her powers or duties under the 1971 Act (A);
Section 19(6)	(o)	without reasonable excuse failing to comply with any requirements of an inspector to give him or her such facilities and assistance to which the defendant's responsibilities extend (whether owner, occupier or employee) as enable him or her to exercise his or her powers (A);
Section 21(1)	(p)	being or having been an inspector and disclosing information other than as authorised under the 1971 Act (A);
Section 22(1)	(q)	knowingly or recklessly giving false information in purported compliance with an obligation under the 1971 Act (B);
Section 22(1)	(r)	with intent to deceive, making or being in possession of a document so closely resembling a fire certificate as to be calculated to deceive (B);
Section 22(1)	(s)	knowingly or recklessly giving false information to procure a fire certificate (B);
Section 22(1)	(t)	making an entry in a book or document required to be kept under the 1971 Act, knowing it to be false (B);
Section 22(2)	(u)	pretending to be an inspector under the 1971 Act, with intent to deceive (A).

Offences by bodies corporate

Section 23(1)

6.8 Where an offence is committed by a body corporate with the consent, connivance or by the neglect of any director, manager, secretary or other similar officer of the body or any person purporting to act in such capacity, that person is also guilty of the offence.

Section 23(2)

6.9 Where the affairs of a body corporate are managed by its members, any member whose acts or defaults in connection with his or her functions of management result in the commission of an offence is guilty of an offence as if the member were a director.

Offences due to the fault of another person

Section 24

6.10 Where the commission of an offence by any person is due to the act or default of some other person, that other person may be charged jointly or separately with the offence and is liable to the same penalty.

Section 25

6.11 It is a defence for the person charged to prove that he or she took all reasonable precautions and exercised all due diligence to avoid the commission of any of the above offences.

Chapter 7: RIGHTS OF APPEAL AND GRIEVANCES

7.1 The Fire Precautions Act 1971 provides for a right of appeal to the courts in certain cases against a decision of the fire authority.

7.2 In the following cases a person may appeal to a magistrates' court (in England and Wales) or to the sheriff (in Scotland) within 21 days from the date on which the decision is made known to that person:-

Section 9(1)   (a) if the appellant objects to any step contained in a notice of steps which he or she is required to take as a condition of issuing or amending a fire certificate;

Section 9(1)   (b) if the appellant objects to any direction given to him or her in connection with any proposed alterations;

Section 9(1)   (c) if the appellant considers that any period of time allowed for taking steps is insufficient and has failed to persuade the fire authority to grant an extension or the extension is insufficient;

Section 9(1)   (d) if the fire authority refuse to issue a fire certificate;

Section 9(1)   (e) if the appellant objects to the inclusion or omission of anything in or from a fire certificate;

Section 9(1)   (f) if the fire authority refuse to amend a fire certificate;

Section 9(1)   (g) if the fire authority refuse to cancel a fire certificate;

Section 9E   (h) if the fire authority issue an improvement notice; or

Section 10A   (i) if the fire authority issue a prohibition notice.

Section 9(1)   **7.3** On hearing an appeal with regard to 7.2(a) to 7.2(g) above the court will make such order as it thinks fit.

Section 9E(2)
Section 10A(2)   **7.4** In the case of an appeal against an improvement or prohibition notice the court may cancel the notice, or affirm it either in its original form or with such modifications as it thinks fit.

Section 27(1)   **7.5** If the appellant, or the fire authority or local authority, is aggrieved by the outcome of the appeal at this stage, there is a further right of appeal to the Crown Court (in England and Wales) or the Sheriff Principal or Court of Session (in Scotland).

Section 9(3)   **7.6** Where an appeal is brought against the refusal of a fire authority to issue a fire certificate with respect to any premises or the cancellation or amendment in pursuance of section 8(7) or 8(9) of the Act of a fire certificate issued with respect to any premises, a person shall not be guilty of an offence under section 7(1) or (2) of the Act by reason of the premises being put to a designated use or used as a dwelling at a time between the relevant date and the final determination of the appeal.

Section 9(4) **7.7** Similarly, if an appeal has been lodged against a requirement imposed in a fire certificate, it is lawful for the appellant not to observe that requirement pending final determination of the appeal.

Section 9E(3) **7.8** If an appeal has been lodged against an improvement notice, this has the effect of suspending the operation of the notice until the appeal is either withdrawn or finally determined.

Section 10A(3) **7.9** However, it is important to note that in the case of a prohibition notice the lodging of the appeal does NOT have the effect of suspending the notice. It will be necessary to apply to the court for the order to be suspended, and if the court so directs then the suspension will have effect from the moment of that direction.

Chapter 8: EFFECT OF THE FIRE PRECAUTIONS ACT 1971 ON OTHER LEGISLATION

Section 30(2)

8.1 Some premises within the scope of the 1971 Act are already covered by provisions in local Acts relating to means of escape and fire precautions. These provisions cease to have effect when such matters are covered by a fire certificate issued in respect of the premises or by regulations made under section 12 of the 1971 Act.

Section 31(1)(a)

Schedule 2: paragraph 7

8.2 Some of the premises in question may be required to be licensed and fire precautions may be imposed as conditions of the licence. Except in the case of premises licensed under the Explosives Act 1875 or the Petroleum (Consolidation) Act 1928, any condition, term etc., imposed by way of licence and which could be dealt with in a fire certificate will be of no effect whilst a fire certificate covering the use, by reason of which the licence was issued, is in force.

Section 32

8.3 A person is excused from doing anything under a local Act which would involve them in a contravention of the 1971 Act or regulations made under it.

Chapter 9: CROWN PREMISES

Section 40

9.1 Section 40 of the Fire Precautions Act 1971 provides for the application of provisions of the Act to Crown premises, ie those premises which are owned and/or occupied by the Crown. Enforcement of the Act including the issue of fire certificates in these premises is the responsibility of fire inspectors (who are members of the Fire Service Inspectorate of the Home Office or The Scottish Office and not the fire authority).

9.2 Where parts of premises are occupied by the Crown but the building is not owned by the Crown, enforcement in those parts which are privately occupied will be the responsibility of the fire authority for the area and in those parts occupied by the Crown will be the responsibility of the Fire Service Inspectorate of the Home Office and The Scottish Office. In such premises it will be normal practice for the fire authority and fire inspector to jointly inspect the premises to ensure that the fire certificates each issue are compatible.

Chapter 10: **NEW BUILDINGS, STRUCTURAL ALTERATIONS AND THE EFFECT OF THE BUILDING REGULATIONS**

Section 13

10.1 In England and Wales, if the premises are in a building to which, at the time it was built or altered, Building Regulations imposing requirements about means of escape in case of fire applied and required plans to be deposited with the local authority, the fire authority may not include in a notice of steps to be taken a requirement to carry out structural or other alterations to the means of escape unless:-

(a) the alterations to the means of escape are necessary to ensure that the premises comply with means of escape requirements of any regulations made under section 12 of the Fire Precautions Act 1971; or

(b) the fire authority are satisfied that the means of escape in case of fire are inadequate by reason of circumstances, particulars of which were not required to be supplied to the local authority in connection with the deposit of plans under Building Regulations.

Note: it is important that the relationship between structural fire precautions and means of escape in case of fire required in respect of the erection, extension or material alteration of a building and the means of escape requirements necessary to satisfy a fire authority for the purpose of issuing a fire certificate are fully understood.

Clearly it would be undesirable if an occupier was required to carry out additional work or incur additional expense to satisfy a fire authority on the matter of means of escape in case of fire if approval for such matters had been obtained under the Building Regulations.

Generally, a fire authority should not need to require any additional steps to be taken in respect of means of escape if the premises have been subject to such requirements in relation to its use under the Building Regulations. Indeed, in all but certain specified circumstances they are statutorily barred by sections 13 and 14 of the Fire Precautions Act 1971 from doing so.

Only if a fire authority is satisfied that the means of escape in case of fire are inadequate by virtue of matters or circumstances that were not required to be shown in connection with the Building Regulations application can they legally require structural or other alterations relating to means of escape, unless such alterations are necessary to satisfy any regulations made under section 12 of the 1971 Act.

The most likely situation that would necessitate further steps to be taken is that of high risk factory premises where the particular process giving rise to the risk was not known or not made known to the Building Control Authority at the time when Building Regulation approval was sought and obtained. There may also occasionally be circumstances where an application for a fire certificate is for a use of premises other than that for which Building Regulation approval was obtained.

It should be understood that the statutory bar applies only where the relevant building is a building on which the Building Regulations imposed requirements as to the means of escape in case of fire.

Section 14

10.2 In Scotland, the fire authority may not require alterations to be made to a building which would bring it up to a standard higher than that required by the Building Standards (Scotland) Regulations, except in certain specified circumstances.

Chapter 11: CONSULTATION

Sections 16 & 17

11.1 The 1971 Act requires consultation between the fire authority and the local authority responsible for the Building Regulations. In England and Wales the local authority are required to consult in specified circumstances the fire authority before dispensing with requirements relating to structural fire precautions or means of escape in the Building Regulations, and before passing plans under them for a building if its first use is likely to be a designated use under the Act. (In Scotland, an applicant wishing relaxation of provisions of Building Standards Regulations must seek a direction from a local authority or The Scottish Office Building Directorate, who consult the Firemaster of the fire authority if requirements for means of escape or structural fire precautions are involved.) The fire authority, for their part, are required to consult the local authority (the local building control authority in Scotland) before requiring alterations to be made to a building in connection with the issue of a fire certificate. These measures are all designed to ensure that there is no avoidable conflict between the requirements of a fire authority and a local authority in respect of fire precautions in the same premises.

Note: Specific advice on consultation in England and Wales between fire authorities and the Department of the Environment was published in the joint Department of the Environment, Home Office, Welsh Office publication "Building Regulation and Fire Safety" Procedural guidance. The document is available from Building Control Departments or Citizens Advice Bureaux.

Section 17(1)(iii)

11.2 There is also a requirement for consultation between the fire authority and the authority enforcing Part I of the Health and Safety at Work etc. Act 1974 (either the Health and Safety Executive or local authorities) before requiring alterations to premises used as a place of work in case their requirements conflict with those of the 1974 Act. There is a corresponding requirement for the enforcing authority under the 1974 Act to consult the fire authority before issuing an improvement or prohibition notice if that notice might lead to the taking of measures affecting the means of escape in case of fire.

PART II

A TECHNICAL GUIDE TO FIRE PRECAUTIONS IN FACTORIES, OFFICES, SHOPS AND RAILWAY PREMISES

DEFINITIONS OF TECHNICAL TERMS USED IN THE GUIDE

Various terms are used in this guide which, because of their importance in regard to **means of escape** for *existing* buildings are defined below. (Defined terms appear in bold type in Parts II and III of the text.)

*Note: It should be noted that the technical terms used in this guide are for **existing** factories, offices, shops and railway premises and may differ from those expressed in Building Regulations which apply to new buildings and those materially altered.*

Access room means a room which forms the only escape route from an **inner room**.

Accommodation stairway means a stairway which is provided for the convenience of occupants additional to that or those required for escape purposes.

Cavity barrier means a construction provided to close a concealed space against penetration of smoke or flame, or provided to restrict the movement of smoke or flame within such a space.

Compartment is a part of the building separated from all other parts of the same building by **compartment walls** and/or **compartment floors.**

Compartment wall/floor means a **fire-resisting** wall or floor used in the separation of one fire **compartment** from another.

Dead-end means an area from which escape is possible in one direction only.

Distance of travel means the actual distance that a person needs to travel between any point in a building and the nearest **storey exit**.

Emergency escape lighting means that part of the emergency lighting system provided for use when the supply to the normal lighting fails so as to ensure that the **means of escape** can be safely and effectively used at all material times.

Final exit means the termination of an escape route from a building giving direct access to a **place of safety** such as a street, passageway, walkway or open space, and sited to ensure that persons can disperse safely from the vicinity of the building and the effects of fire.

Fire door means a door assembly which if tested under:-

(a) the conditions of test for door assemblies described in British Standard 476: Part 22; or

(b) the conditions of test contained in the British Standard currently in force at the time of the bringing into use of the premises as a factory, office, shop or railway premises; or

(c) the conditions of test in the British Standard currently in force at the time the door was manufactured;

would satisfy the criteria for integrity for 20 minutes or for such longer period as may be specified in particular circumstances.

Notes: 1. Normally such doors should be positively self-closing.

2. Door assemblies with non-metallic leaves should be maintained in accordance with the provisions of British Standard 8214.

Fire-resisting (fire resistance) means the ability of a component or construction of a building to satisfy for a stated period of time some or all of the appropriate criteria specified in the relevant part of British Standard 476.

*Note: In **existing** premises, it may not be possible to confirm the **fire resistance** of some elements of structure and a judgement will need to be made on whether the **fire resistance** is acceptable.*

Fire safety engineering is an approach which takes into account the total fire safety package and sets a range of fire safety features against an assessment of the fire hazard and fire risk for the particular premises.

Fire/smoke stopping is a seal provided to close an imperfection of fit or design tolerance between elements or components to restrict the passage of fire, heat and smoke.

Inner room means a room from which escape is possible only by passing through an **access room.**

Means of escape is the structural means whereby a safe route is provided for persons to travel from any point in a building to a **place of safety** beyond the building without outside assistance.

Place of safety means a place beyond the building in which a person is no longer in danger from fire.

Protected corridor means a corridor which is adequately protected from fire in adjoining accommodation by **fire-resisting** construction.

Protected lobby is a **fire-resisting** enclosure providing access to a **protected stairway** via two sets of **fire-resisting** self-closing doors and into which no rooms open, other than toilets or lifts.

Protected route means a route having an adequate degree of protection from fire including walls (other than any part that is an external wall of a building), doors, partitions, ceilings and floors separating the route from the remainder of the building.

Protected stairway means a stairway which is adequately protected from fire in adjoining accommodation by **fire-resisting** construction and either discharges through a **final exit** or a **protected route** leading to a **final exit**.

Storey exit means an exit through which persons are no longer at immediate risk from the effect of fire and includes a **final exit**, an exit to a **protected lobby** or **protected stairway** (including an exit leading on to an external stairway), and an exit provided for **means of escape** through a **compartment wall** via which a **final exit** can be reached.

Chapter 12: FIRE RESISTANCE AND SURFACE FINISHES OF WALLS, CEILINGS AND ESCAPE ROUTES

Fire resistance

12.1 When planning fire precautions and **means of escape** in premises it is usual to have regard to the **fire resistance** of the elements of structure, eg walls, floors, etc. In the types of premises covered by this guide it may not always be possible to achieve the minimum standards set out in Table A. In such circumstances compensating features are required such as a reduction in the **distance of travel** or the provision of other fire safety installations (see paragraph 13.3 and paragraphs 15.12 to 15.14).

12.2 The period of **fire resistance** recommended in Table A should be achieved wherever practicable. Where the main elements of structure offer less than a 60 minute standard, an enclosure to an area of high fire risk and the floor immediately over a basement should, *under no circumstances*, be less than 30 minutes **fire resistance.**

Table A Minimum fire resistance (integrity in minutes)
This table should be read in conjunction with the Notes which follow the table.
(Figures in brackets refer to the accompanying notes)

	Offices and Shops			Factories		
	Walls	Doors	Floors	Walls	Doors	Floors
Enclosing an area of high fire risk	60 (8 & 9)	60 (7, 8 & 9)	60 (8, 10 & 11)	60 (8 & 9)	60 (7, 8 & 9)	60 (8, 10 & 11)
Enclosing a **protected route**	30 (9)	30 (3, 5, 7 & 9)	30	30 (9)	30 (3, 5, 7 & 9)	30
Enclosing a stairway	30 (9)	30 (3, 5, 7 & 9)	30	30 (9)	30 (3, 5, 7 & 9)	30
In a stairway from ground floor to basement	30 (9)	2 × 30 (5, 7 & 9)	—	60 (10)	2 × 30 or 1 × 60 (5, 6, 7 & 9)	—
In a corridor solely to sub-divide it or at its junction to separate a **dead-end**	—	20 (5, 7 & 9)	—	—	20 (5, 7 & 9)	—
Enclosing a **compartment**	30 (9)	30 (5, 7 & 9)	30	60 (9)	60 (7 & 9)	60 (10)
Enclosing a lift well	30	30 (4, 5 & 7)	30	30	30 (4, 5 & 7)	30
Enclosing a lift motor room	30 (9)	30 (5, 7 & 9)	30	30 (9)	30 (5, 7 & 9)	30
Enclosing a ventilation duct	30 (1)	30	30	30 (1)	30	30
Floor immediately over a basement	—	—	30 (1, 2 & 10)	—	—	60 (1, 2 & 10)
All other floors	—	—	30 (1 & 2)	—	—	30 (1 & 2)

NB: In railway premises the minimum **fire resistance** for offices and shops will generally apply.

Notes to Table A

1. *See paragraph 12.7.*

2. *This does not include incomplete floors such as gallery floors or raised floor areas.*

3. *Except a door to a toilet containing no fire risk, provided that the toilet room is separated by **fire-resisting** construction from the remainder of the building.*

4. *Except a lift well contained within a **protected stairway** enclosure (see also paragraphs 14.72 and 14.73.)*

5. *An existing timber door may be deemed to satisfy the necessary standard of **fire resistance** if it can be suitably modified either in accordance with the methods recommended in the Timber Research and Development Association's wood information sheet, section 1, sheet 32 "Fire resisting doorset by upgrading", or by following other suitable methods.*

6. *See paragraph 14.53.*

7. *See paragraphs 14.31 and 14.83(b)iv; and the technical definition of "**Fire Door**". Doors to corridors forming a **dead-end** and those doors separating the **dead-end** where it joins a main corridor (see diagram 4), should have 20 minutes integrity unless they are affording protection to a stairway.*

8. *Examples of these high fire risk areas are where highly flammable materials may be stored or where the processes or activities may give rise to the risk of rapid fire development (see paragraphs 13.8 to 13.13).*

9. *(a) **Fire-resisting** glazed areas may be incorporated into the wall separating any accommodation from a **protected corridor** provided that the glazed element would satisfy the **fire resistance** criterion for the period given in the table were it tested to British Standard 476: Part 22.*

   *(b) A glazed screen may be used to separate an escape corridor from a stairway enclosure provided that the appropriate **fire resistance** given in the table is maintained between the two areas. The insulation criterion of British Standard 476 may be waived where the glazing does not extend below 1.1 metres above the adjoining floor level. However, where wheelchair users may use the route the bottom edge of glazing in a door should be lowered to 900mm above floor level.*

   *(c) Vision panels incorporated into a door which protects an escape route should not reduce the **fire resistance** required for the door and should accord with the appropriate provisions of British Standard 6262.*

10. *See paragraph 12.2.*

11. *Other than ground floor.*

Surface finishes of walls, ceilings and escape routes

12.3 Where buildings have been subject to requirements relating to internal fire spread (linings), in connection with the depositing of plans with the local authority for the purpose of Building Regulations, the surface finishes of walls and ceilings should be maintained to the standard in respect of which Building Regulation approval was obtained.

12.4 Materials used to line walls and ceilings may be difficult to assess in terms of their contribution to the spread of flame and the development of fires. The following are examples of the type of finishes which should meet the standards set out in Table B. Where there is doubt, as to whether a manufactured or treated surface meets the appropriate standard, written evidence of the standard achieved should be obtained from the manufacturer or supplier. Suitable maintenance of any special coatings should be provided for in the requirements imposed by the fire certificate.

Table B **Minimum classes for surface spread of flame**

Class 0	In circulation spaces and escape routes
Class 1	In rooms, (other than small rooms) and places of assembly
Class 3	In small rooms (see example for Class 3)

Examples

Class 0: Acceptable in all locations including circulation spaces and escape routes.

Brickwork, blockwork, concrete, plasterboard, ceramic tiles, plaster finishes (including rendering on wood or metal laths), woodwool slab, thin vinyl and paper coverings on inorganic surface (other than heavy flock wallpapers) and certain thermosetting plastics.

Class 1: Acceptable in all rooms but not acceptable on escape routes such as stairways, corridors, entrance halls.

Timber, hardboard, blockboard, particleboard (chipboard), heavy flock wall papers, thermosetting plastics, that have been flame retardant treated to achieve a Class 1 standard.

Class 3: Acceptable in small rooms (ie not exceeding 30m^2) and on parts of the walls of other rooms if the total area of those parts does not exceed an area equivalent to one half of the floor area subject to a maximum of 60m^2. Not acceptable on escape routes such as stairways, corridors, entrance halls or in rooms other than as specified above.

Timber, hardboard, blockboard, particleboard (chipboard), some heavy flock wall papers, thermosetting plastics and thermoplastics (expanded polystyrene wall and ceiling linings).

Notes: *1. Classes 1 and 3 are classifications determined by reference to a test method specified in British Standard 476: Part 7, Class 1 being the best.*

2. The classification Class 0 is not referred to in a British Standard test but refers to a standard which restricts both the spread of flame across a surface and also the rate at which heat is released from it. It is a higher standard than Class 1 and is referred to in Approved Document B which gives guidance on the way the functional requirements of Part B of the Building Regulations may be met in England and Wales. It is also specified as the prescribed or deemed-to-satisfy standard in Part E of the Technical Standards associated with the Building Standards (Scotland) Regulations.

12.5 Partitions, space dividers and other similar vertical surfaces which are provided to sub-divide a room should not be less than the class of surface required for the room in which they are situated.

Display materials

12.6 Any display comprising large amounts of paper, textiles or flimsy material, particularly in circulation areas, can cause fire to spread rapidly and negate the advantages of suitable wall and ceiling linings. Cotton wool and cellular material particularly should be avoided and it is also important that displays are confined to appropriately located display boards.

Fire/smoke spread

12.7 Fire, heat and smoke can spread by way of:-

(a) service openings, eg ductwork, pipework openings, chutes and ventilation trunking; or

(b) horizontal or vertical voids between floors and ceilings.

It will therefore be necessary to safeguard the **means of escape** by providing such measures as **fire/smoke stopping, cavity barriers**, and fire dampers within ductwork. This is particularly important where services penetrate **fire-resisting** floors or walls. Guidance on **fire/smoke stopping** in **cavity barriers** is given in Approved Document B of the Building Regulations and the Technical Standards to the Building Standards (Scotland) Regulations 1990. Recommendations on the provision of fire dampers are contained in British Standard 5588: Part 9. (See note to paragraph 14.31.)

12.8 As voids and wall cavities may not be readily apparent it will be necessary to carry out a detailed examination of structural separation. Without such an inspection and any identified defects being remedied, it is possible that fire, heat and smoke could pass unrestricted through these openings, thereby jeopardising the use of escape routes.

Notes/Amendments

Chapter 13: ASSESSMENT OF FIRE RISK AND ASSOCIATED LIFE RISK

Introduction

13.1 As premises covered by this guide can vary considerably in size, layout and construction, the risk of a fire occurring can also vary, much depending on the work activity. For instance the risk may be greatly increased in factories where hazardous substances are stored or used. It is essential, therefore, that the fire safety measures are determined having regard to all relevant circumstances.

13.2 It is not possible to offer clear-cut, or hard and fast guidance for making these assessments because all factors have to be taken into account but paragraphs 13.5 to 13.13 describe in general terms the issues which will need to be considered to determine the level of fire risk (ie low, normal or high) thus enabling adoption of the appropriate fire precautions.

13.3 The details contained in the following paragraphs should be treated as broad indicators. It does not necessarily follow that the presence (or indeed the absence) of one of the factors mentioned in the description of risk category means that the premises or part of the premises should be placed in the low or high categories. It is likely that in many premises there will be a variety of risks and it is important that all factors are considered, including any automatic fire detection and/or suppression system which has been installed. The presence of the latter eg sprinklers installed for the overall protection of the building, or other fire extinguishing systems covering areas of high fire risk, may significantly reduce the dangers of rapid fire growth and consequently will have a bearing on the final risk assessment.

Note: Section 15 of the Fire Safety and Safety of Places of Sport Act 1987 came into force on 1 August 1993 (see paragraph 13 of the Introduction).

13.4 In certain premises, such as buildings containing atria and in town centre developments, the design features associated with the construction and layout of such premises and developments may present special and unusual difficulties in so far as the danger of fire and smoke spread is concerned. Recommendations dealing with these and other matters in connection with fire precautions in town centre developments can be found in British Standard 5588: Part 10. The manner of assessing the fire risk in these particular instances will, however, generally be in accordance with that given in paragraphs 13.5 to 13.8.

Assessment of low fire risk

13.5 Low fire risk premises are those where there is minimal risk to life safety and where the risk of fire occurring is low, or the potential for fire, heat and smoke spreading is negligible. Such premises include those used for heavy engineering or where the process if entirely a wet one and non-combustible materials predominate.

Assessment of normal fire risk

13.6 Normal fire risk premises are those where:-

(a) any outbreak of fire is likely to remain confined or is likely to spread only slowly, thereby allowing persons time to escape to a **place of safety**; and/or

(b) the presence within the premises of an effective automatic means for giving warning in the case of fire, or an effective automatic fire extinguishment, suppression or containment system may reduce the fire risk classification from high fire risk.

13.7 Premises within the normal fire risk category will also be those where the use of the building as well as its contents are unlikely to present a serious risk to the occupants in the event of fire, eg offices and shops selling goods which are not easily ignited.

Assessment of high fire risk

13.8 Factors which lead to the assessment of premises or parts of the premises as being of high fire risk include the following:-

(a) the presence of highly flammable or explosive materials (other than in small quantities);

(b) the presence of unsatisfactory structural features which may promote the spread of fire, heat and smoke; and

(c) the permanent work activity or temporary work activity particularly work using heat producing processes (see paragraph 13.11).

The presence of highly flammable or explosive materials (other than in small quantities)

13.9 In such cases, the materials will either be easily ignited or likely, when ignited, to cause the rapid spread of fire, heat or smoke. The materials may be solid eg man made textiles such as acrylics, or may be present as liquid, spray, vapour or dust; in general such materials are more likely to be present in factories than in other occupancies (either in manufacturing processes or in storage). However, it is not uncommon to find similar materials in some shops, particularly in shops retailing highly combustible products (or products having highly combustible constituent parts); and in these circumstances the risk may be made greater by any concentration of these products. Some furniture, furnishings and decorative materials are also combustible (see Chapter 20).

The presence of unsatisfactory structural features which may promote the spread of fire, heat and smoke

13.10 Such features include:-

(a) a lack of **fire-resisting** separation:

(b) vertical or horizontal openings through which fire can spread, eg a goods lift connecting work areas, service ducting or conveyor belts, **accommodation stairways** and false floors, ceilings and voids;

(c) wooden floors supported on wooden joists, particularly if soaked in oil or other combustible liquid eg workshops, factory warehouses in dock areas, mill premises;

(d) long and complex escape routes caused by extensive sub-division of large floor areas by partitions, or the distribution of display units or goods in shops or machinery in factories; and

(e) large areas of flammable or smoke producing surfaces on either walls or ceilings eg combustible wall board, heavy flock wallpaper, vinyl or plastic; and surfaces covered by a build-up of flammable paint or other decorative surfaces.

The permanent work activity or temporary work activity, particularly work using heat producing processes

13.11 Such activities and areas include:-

(a) workshops in which manufacturing processes involve the use of highly flammable liquids eg paint spraying;

(b) areas where the processes involve the use of naked flame, or produce excessive heat;

(c) areas which may contain different chemicals either stored or used which, if not strictly managed, may become violently reactive, or give off flammable gas or vapours;

(d) large kitchens associated with restaurants (public or staff);

(e) heating systems and their fuel storage areas/containers inadequately separated from the remainder of the premises;

(f) areas where explosive or highly flammable materials are stored or used;

(g) areas where upholstered furniture is stored; and

(h) refuse chambers and waste disposal areas.

Note: In relation to (a), (b) and (c) see also paragraph 10 of the Introduction.

Other considerations

13.12 The occupants of the premises (employees, visitors, or members of the public) can also have a bearing on the overall risk assessment in relation to life safety. These include:-

(a) individuals or small groups of persons working in isolated parts of the building such as those working in basements, roof spaces, cable ducts etc;

(b) large numbers of persons present relative to the size of the building (eg sales at department stores), so that only a low level of assistance is available to the public in an emergency; and

(c) a high proportion of the elderly, or people with temporary or permanent disabilities.

Storage and display in shops

13.13 In certain areas, because of the nature of the materials displayed or stored or because of the methods of display or storage, the use of the standards set out in this guide will not be sufficient to ensure evacuation within a reasonable time. This situation is particularly likely where large quantities of readily ignitable and highly flammable materials are stored or displayed. In such circumstances special consideration should be given to the adoption of the following remedial measures:-

(a) not allowing goods to be stored in those parts of the premises which are sales areas ie areas to which the public normally have access;

(b) restricting the storage of goods to those parts of the premises to which the public are not admitted. Such parts should be separated from the remainder of the premises by an enclosure, the floors, walls and self- closing doors of which have a standard of **fire resistance** of not less than 60 minutes. (See paragraph 12.1). The enclosure should not form any part of an escape route which would have to be used by the public in case of fire;

(c) ensuring that the storage of goods is arranged so that there is a clear passageway from any point in the storage enclosure to the **means of escape**;

(d) ensuring that displays of goods in public areas are so arranged that there is unimpeded access to all gangways leading to exits and that the display does not obstruct a clear view of exits and their associated exit signs;

(e) making sure that, wherever practicable, the display of goods is not:-

    (i) on the same floor as an area which will attract large numbers of the public at any one time eg a restaurant or licensed bar; or

    (ii) in any part where persons are invited to wait and receive specialist services eg hairdressing and beauty treatment salons;

unless there are adequate **means of escape** to a **place of safety** which in (i) does not involve passing through an area where goods are displayed.

Note: Cellular foam filled furniture presents particular dangers if it is involved in a fire. Any display of such furniture in public areas should be kept to the minimum necessary for daily trading needs and the overall guidance given in paragraph 13.13 should be carefully followed.

(f) ensuring that goods on display which are accessible to the public eg for demonstration or examination purposes, are not stacked in a manner which increases the fire hazard, nor covered with materials such as dust covers which can be easily ignited;

(g) encouraging managements to:-

    (i) prohibit smoking in all areas where combustible goods are stored or displayed; and

    (ii) provide conspicuous notices to that effect.

(h) if the occupier proposes to instal a sprinkler system, this should be installed and maintained in conformity with British Standard 5306: Part 2, ideally including additional features required for life safety. Any sprinkler system or other automatic fire suppression or detection installation should be linked into the fire warning system for the building.

Notes/Amendments

Chapter 14: MEANS OF ESCAPE

Introduction

14.1 This chapter deals with the **means of escape** and provides guidance for *existing* factories, offices, shops and railway premises. Because of the range of premises to which this guide applies a flexible approach needs to be adopted with account being taken of the basic principles described in paragraphs 14.3 to 14.8. Existing buildings may also not have the structural fire precautions currently required by Building Regulations and consequently, it may be decided that some of the standards, particularly the **distances of travel**, should be more onerous than for new buildings. However, where a building has been the subject of **means of escape** requirements under Building Regulations (see Chapter 10), and been maintained to those standards this chapter of the guide will generally not apply.

14.2 Basic standards relating to **means of escape** are set out in the following paragraphs which also contain simple diagrams (not to scale) to show some of the principles which should be applied.

Note: It is emphasised that the diagrams provided are only examples of the principles to be applied.

General principles

14.3 The principle on which **means of escape** provisions are based, is that persons, regardless of the location of the fire, should be able to proceed safely along a recognisable escape route by their own unaided efforts. However, it may be necessary for the occupier/owner to make additional provisions for people with disabilities. British Standard 5588: Part 8: Code of Practice for means of escape for disabled people gives guidance (see also Chapter 19).

14.4 Other basic principles for **means of escape** include the following:-

(a) there should be alternative **means of escape** from most situations unless the **distance of travel** for one direction only complies with Table D;

(b) the distance that persons should travel to reach a **place of safety** will be dependent upon the fire risk assessment for the premises – the greater the risk, the shorter the acceptable **distance of travel**;

(c) where direct escape to a **final exit** is not possible, a place of comparative safety, such as a **protected stairway**, should be reached within a reasonable **distance of travel** (see paragraph 14.13 and Table D);

(d) unless a **place of safety** can be reached within a reasonable **distance of travel** (see Table D) the escape routes will need to be protected from the effects of fire elsewhere in the building (eg by **fire-resisting** construction);

(e) escape routes should always terminate in a **place of safety**;

(f) escape routes should be wide enough to cater for the number of occupants likely to use them and should not reduce in width;

(g) there should be a sufficient number of available exits of adequate width from a room, storey or building; and

(h) the required exits should be so spaced that persons can turn their backs on a fire and proceed in the opposite direction to a **place of safety**.

*Note: A **protected route** may consist of two interlinked structural components -*

*(a) a **protected corridor** (horizontal movement); and*

*(b) a **protected stairway** (vertical movement);*

although in some situations only one of these components may be present.

*The principle to be followed is that once persons enter either (a) or (b) they should normally be able to proceed to a **place of safety** without leaving the **protected route**. Exceptions to this general principle are shown in diagrams 4 and 14.*

*The protection required at (a) and (b) is generally achieved by ensuring that the protected area has at least 30 minutes **fire resistance**.*

14.5 When the minimum number of exits have been calculated, their location should also be considered to ensure that the exits will fulfil their purpose in an emergency. For example, there could be in a particular building, a sufficient number of exits all of which could be sited at one end of the occupied area. A fire in this vicinity could conceivably prevent access being gained to any or all of the exits. A number of exits which discharge into a common space cannot be regarded as alternatives to each other.

14.6 Certain features are not acceptable as a **means of escape** in case of fire. Examples of these are:-

(a) lifts (except where specifically designed or adapted for the evacuation of people with disabilities in accordance with British Standard 5588: Part 8);

(b) escalators (but see paragraphs 14.57 to 14.59);

(c) portable, foldaway, cantilever or throw-out ladders (but see paragraph 14.67); and

(d) lowering lines and other self rescue devices (but see paragraph 14.69).

14.7 Items which pose a potential fire hazard and those which could cause an obstruction should not be installed or stored in corridors and stairway enclosures. The following are examples of such items and *under no circumstances* should such items be allowed where the stairway is the only **means of escape**. Examples of items are as follows:-

(a) portable heaters of any type;

(b) heaters which have unprotected naked flames or radiant bars;

(c) fixed heaters using a gas supply cylinder where the cylinder is within the escape route;

(d) oil-fuelled heaters or boilers;

(e) cooking appliances;

(f) stored furniture;

(g) temporarily stored items including items in transit (eg furniture, items of stationery, display goods etc.);

(h) lighting using naked flames;

(i) gas boilers, pipes, meters or other fittings (other than those installed in accordance with appropriate Gas Safety Regulations);

(j) coat racks;

(k) photocopying, gaming or vending machines; and

(l) electrical equipment other than normal lighting, **emergency escape lighting**, fire alarm systems or security equipment.

Alternative routes from a room or storey

14.8 Alternative escape routes from a room or storey should generally satisfy the following criteria:-

(a) they are in directions 45° or more apart (see diagram 1); or

(b) from any point from which there is initially a single direction of escape they diverge by 45° plus 2.5° for every metre travelled in one direction (see diagram 2).

Diagram 1: Example of siting of exits to show the 45° rule.

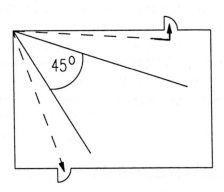

Diagram 2: Example of siting of exits to show the angle of divergence within a room in a normal fire risk shop.

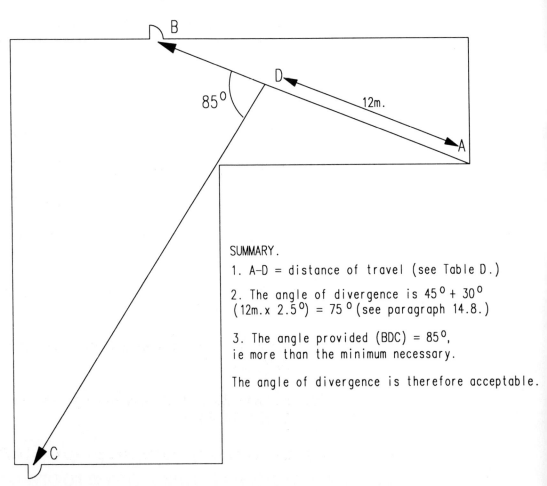

SUMMARY.
1. A–D = distance of travel (see Table D.)
2. The angle of divergence is 45° + 30° (12m. x 2.5°) = 75° (see paragraph 14.8.)
3. The angle provided (BDC) = 85°, ie more than the minimum necessary.

The angle of divergence is therefore acceptable.

Fire safety engineering

14.9 Where it is impracticable to achieve adequate **means of escape** in accordance with the principles described in this chapter, it may be possible, in certain circumstances, to accept a **fire safety engineering** solution.

14.10 Fire safety engineering is a package of fire safety arrangements which offers an equivalent standard of safety to that which would be achieved by applying the conventional methods of structural fire protection. For example, the provision of a smoke management system or a sprinkler system may allow a reduction in either the standard of **fire resistance** required or the acceptance, in certain circumstances, of extended travel distances. The use of pressurization in a stairway may similarly allow acceptance of a lower standard of **fire resistance** to doorways entering on to that stairway.

Assessment of means of escape

14.11 A number of factors which are inter-related need to be taken into consideration when assessing **means of escape**. Such matters include:-

(a) assessment of fire risk (see Chapter 13);

(b) **fire-resisting** construction, fire/smoke spread and the surface finishes of walls, ceilings and escape routes (see Chapter 12 and Tables A and B);

(c) occupant capacity (see paragraph 14.12 and Table C);

(d) the ability of occupants to respond (see Chapter 19);

(e) **distance of travel** (see paragraphs 14.13 to 14.55 and Table D);

- Travel within rooms (see paragraphs 14.18 to 14.25);

- Travel from rooms to a **storey exit** (see paragraphs 14.26 to 14.32);

- Travel within stairways and to **final exits** (see paragraphs 14.33 to 14.55);

(f) doors on escape routes and their fastenings (see paragraphs 14.83 to 14.92);

(g) exit and directional signs and signs on doors (see paragraphs 14.93 to 14.99); and

(h) lighting of escape routes (see paragraphs 14.100 to 14.102).

Occupant capacity

14.12 The occupant capacity is the maximum number of persons which can be safely accommodated in a building or a part of a building. This number can be calculated by using the occupant floor space factors in Table C.

Note: Different occupant floor space factors may be needed for different parts of the building depending on the use.

Table C   Occupant floor space factors commonly accepted
This Table should be read in conjunction with the Notes which follow the table.

	Type of accommodation, *(Notes 1, 2 and 7)*	Floor space factor m²/person
i.	Standing spectator areas.	0.3
ii.	Amusement arcade, assembly hall (including a general purpose place of assembly), bingo hall, club, crush hall, dance floor or hall, venue for pop concert and similar events, queuing area. *(Note 4)*	0.5
iii.	Bar.	*0.3–0.5
iv.	Concourse or shopping mall. *(Note 3)*	0.75
v.	Committee room, common room, conference room, dining room, licensed betting office (public area), lounge (other than a lounge bar), meeting room, reading room, restaurant, staff room, waiting room. *(Note 4)*	*1.0–1.5
vi.	Exhibition hall.	1.5
vii.	Studio (radio, film, television, recording)	1.4
viii.	Shop sales area, *(Note 5)* skating rink.	2.0
ix.	Art gallery, dormitory, factory production area, office (open-plan exceeding 60m²), workshop.	5.0
x.	Kitchen, library, office (other than in ix. above), shop sales area. *(Note 6)*	7.0
xi.	Study bedroom.	8.0
xii.	Bowling alley, billiards or snooker room.	9.3
xiii.	Storage and warehousing.	30.0
xiv.	Car park. 2 persons per parking space	

* depending upon the amount of seating and tables provided.

Notes: 1. In certain circumstances, the floor space factors in Table C may vary slightly from those set out in Approved Document B to the Building Regulations and the Technical Standards to the Building Standards (Scotland) Regulations 1990.

2. Where accommodation is not directly covered by the descriptions given, a reasonable value based on a similar use should be selected.

3. See section 4 of British Standard 5588: Part 10 for detailed guidance on the calculation of occupancy in common public areas in shopping complexes.

4. Alternatively, the occupant floor space factor may be taken as the number of fixed seats provided, if all the occupants will normally be seated.

5. Shops other than those included under item x, but including supermarkets and department stores (all sales areas), shops for personal services such as hairdressing and shops for the delivery or collection of goods for cleaning, repair or other treatment or for members of the public themselves carrying out such cleaning, repair or other treatment.

6. Shops (excluding those in covered shopping complexes and department stores) trading predominantly in furniture, floor coverings, cycles, prams, large domestic appliances or other bulky goods, or trading on a wholesale self-selection basis (cash and carry).

7. Where any room or floor is used or is likely to be used for a variety of purposes, the most appropriate occupant floor space factor should be applied.

Distance of travel

14.13 The factors which have to be considered when assessing **means of escape** will vary widely from one set of premises to another. Accordingly the **distances of travel** suggested in the following paragraphs and Table D *should be regarded as guidelines* and not as hard and fast limits. There are likely to be many situations at both ends of the scale in which either reductions are necessary or increases may be possible. See also paragraphs 14.9 and 14.10 on **fire safety engineering** and Chapter 13 in relation to the assessment of fire risk and associated fire risk.

Table D   Guidelines on distance of travel in metres

This Table should be read in conjunction with the Notes which follow the table.

A: TYPE OF PREMISES	FACTORY			SHOP		OFFICE
B: CATEGORY OF FIRE RISK	1 High	2 Normal	3 Low	4 High	5 Normal (see para 13.7)	6 Normal (see para 13.7)
C: WITHIN A ROOM OR ENCLOSURE: i: In one direction only ii: In more than one direction	6 12	12 25	25 (Note 1)	6 12	12 (Note 1)	12 (Note 1)
D: FROM ANY POINT IN AN **INNER ROOM** TO THE EXIT FROM THE ACCESS ROOM	6	12	25	6	12	12
E: TOTAL **DISTANCE OF TRAVEL** TO **STOREY EXIT**: i: In one direction only ii: In more than one direction	12 25 (Note 2)	25 45	45 60	12 25 (Note 2)	18 (Note 3) 30	25 45

Notes: 1. *Generally the **distances of travel** in line C ii are double those of C i. However in the case of low fire risk factories and normal fire risk offices and shops, the **distances of travel** should be assessed in accordance with the circumstances presented.*

2. *See paragraph 14.16.*

3. *The **distance of travel** for shops in comparison with offices recognises that members of the public will be unfamiliar with the escape routes.*

4. *In railway premises the **distance of travel** selected should be that most appropriate to the use to which the particular part of the premises is being put in accordance with Table D. However, in the case of railway platforms, and concourses at large termini, greatly extended **distances of travel** may well be acceptable. It should be appreciated that these premises are designed for the effective ingress and egress of large numbers of people and are in many cases, for all practical purposes, covered open spaces. Special consideration may need to be given to rafted platforms (ie where the headroom above the platform and track has been reduced by the construction of additional floors) see also Note 5.*

5. *British Rail and the London Fire and Civil Defence Authority have published technical guidance entitled "Guidance for fire precautions on existing British Rail surface stations" which is commended for use as an adjunct to this guide.*

14.14 Where a building is divided into a number of rooms with linking corridors, it will be necessary to consider **means of escape** in the following ways:-

- travel within rooms; and

- travel from rooms to a **storey exit**.

However, in many instances when applying the recommendations on **distances of travel** to reach a **place of safety**, particularly in premises where an open floor area discharges directly to **storey exits**, whether or not these are **final exits**, it may be appropriate to consider only the one distance ie the "total **distance of travel**" to a **storey exit** dealt with in line Ei or Eii of Table D (see paragraph 14.46).

Measuring distance of travel

14.15 **Distance of travel** should be measured as being the actual distance to be travelled between any point in a building and the nearest **storey exit.**

14.16 In cases where more than one element of escape exists or where, for example, part of an escape route is within a room and the remainder is within another adjoining room, the recommended distances should be applied according to the assessment of the risk in each of the rooms. Diagram 3 provides an example of this situation in a factory in which escape is in more than one direction. The first part of the escape route is in a room assessed as high fire risk and the second is within a part judged to be low fire risk. This means that the distance should be limited in the first room to about 12 metres (see line Cii column 1 of Table D) and in the second to 48 metres, ie a total distance to a **storey exit** of 60 metres (see line Eii column 3 of Table D).

Diagram 3   Example of **distance of travel** in a factory with both high and low fire risk areas.

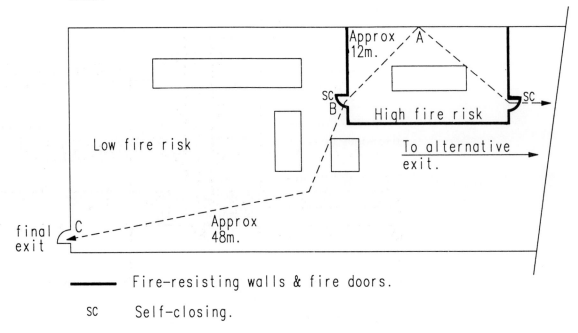

Initial dead-end

14.17 Where an escape route consists initially of a **dead-end** and then has an alternative route to a **storey exit**, the distance in the **dead-end** should not generally exceed the distances shown in line Ei of Table D as appropriate to the location from which it is to be measured; and the total **distance of travel** to a **storey exit** should not normally exceed the distance shown in line Eii of Table D. Diagram 4 is an example of this point for office premises of normal fire risk (see also paragraphs 14.28 and 14.29).

Diagram 4 Example of where the escape route consists initially of one direction only from an office of normal fire risk (see paragraphs 14.17 and 14.29).

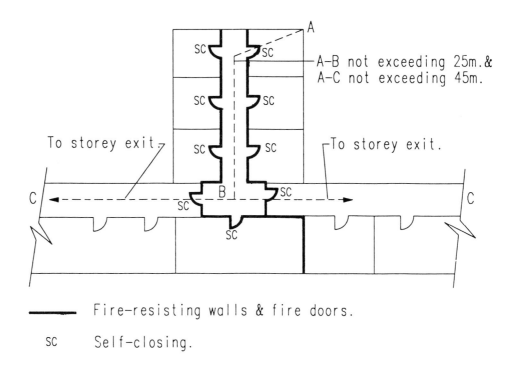

——— Fire-resisting walls & fire doors.

SC   Self-closing.

Travel within rooms

Inner and access rooms (including enclosures)

14.18 Where the only **means of escape** from any room is through another room (an **access room**) the following provisions should apply:-

(a) the number of persons occupying the **inner room** should be kept as small as possible;

(b) the **means of escape** from the **inner room** should not pass through more than one outer room;

(c) the **distance of travel** from any point in the **inner room** to the nearest exit from the **access room** should not exceed that given in Table D;

(d) the **access room** should not be an area of high fire risk and should be in the control of the same occupier as the **inner room**;

(e) in order that the occupants of the **inner room** can be aware of an outbreak of fire in the **access room**, one of the following arrangements should be made:-

(i) the **access room** should be fitted with a suitable automatic fire detection device to give audible warning to the occupants of the **inner room** (see Chapter 15); or

(ii) a vision panel of suitable size should be located in the door or walls of the **inner room** to give clear unobstructed vision into the **access room**; or

(iii) the partition wall separating the **inner room** from the **access room** should be stopped at least 500mm below the ceiling.

Diagram 5 Example of **access room** showing each of the alternative provisions set out in paragraph 14.18(e)(i) to (iii).

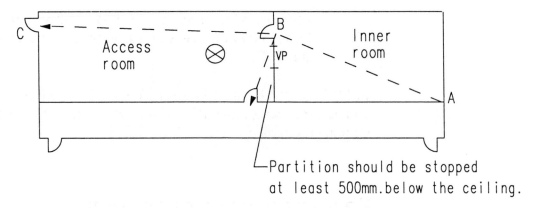

VP — Vision Panel or

⊗ — Automatic fire detection.

14.19 Where the door of an **inner room** opens into an **access room** of low or normal fire risk and between the rooms there is a vision panel (or similar facility) or if automatic fire detection is installed in the **access room** as described in paragraph 14.18, the restriction on **distance of travel** recommended in that paragraph need not be applied if from the point of the exit from the **inner room** there is escape in more than one direction from the **access room** (see also paragraph 14.8).

Number and width of exits from rooms

14.20 More than one exit will be required in the following situations:-

(a) if a room is to be occupied by more than 60 persons; or

(b) if the **distance of travel** between any point and the only exit is more than the appropriate distance recommended in line Ci of Table D.

14.21 There may also be circumstances where the risk of rapid fire spread is so great that, irrespective of the size of the room or the number of persons accommodated, a minimum of two exits are essential eg, factory premises where processes involve spraying with highly flammable liquid.

14.22 The clear width of an exit from any room should not normally be less than 750mm unless the exit will be used by fewer than 5 persons (see also paragraphs 14.80 and 14.82 and Chapter 19).

14.23 Where it is necessary to have more than one exit for **means of escape** purposes, the aggregate capacity of all exits, less the largest of them, should not be less than:-

(a) 750mm for an occupancy of up to 100 persons; and

(b) 1.05 metres for an occupancy of up to 200 persons. An additional 75mm should be allowed for every 15 (or part of 15) persons above 200.

14.24 The width of doorways should be measured as the clear unobstructed width through the doorway where the doors are open at right angles to the frame.

Passageways through rooms

14.25 The contents of any room or **compartment** should be so arranged and maintained as to ensure that there are unobstructed passageways leading to **storey exits**. Often in factories the layout of benches, plant, etc within working areas will be such as to clearly identify passageways. A similar situation will generally occur in offices, particularly those which are open plan. Where this is not the case, and particularly in large shop sales floor areas where the random display of goods could encroach into required passageways, it may be necessary to ensure that routes are clearly defined by, for example, painting lines on the floor or, providing a contrasting floor covering. Where the spaces to be reserved as passageways periodically change, there will be a need to identify these routes by using some form of durable marking.

Travel from rooms to a storey exit

14.26 Escape in more than one direction may be deemed to be available from any point from where there are alternative routes leading to separate **storey exits.**

Corridors

14.27 A main corridor should not normally be less than 1.05 metres wide.

14.28 In corridors where escape is in one direction only, the route should be a **protected route** unless:-

(a) it leads to a **final exit**, within the recommended **distance of travel** to a **storey exit** as shown in line Ei of Table D; and

(b) the fire risk and associated life risk assessment throughout the route is low.

14.29 Where a corridor is required to be a **protected route** and consists initially of a **dead-end** and then has alternative routes, where both sections of the corridor join, they should be separated by **fire-resisting** construction. The distance that a person should travel from any point within the **dead-end** to reach that junction should not exceed the distance recommended in line Ei of Table D (see diagram 4, paragraph 14.17 and Notes 7 and 8 to Table A at paragraph 12.1).

14.30 Where the corridor is a **protected route**, doors to cupboards, service ducts and any vertical shafts linking floors, should be **fire doors** and should be marked "Fire Door – Keep locked shut when not in use". (see paragraph 14.98).

14.31 Corridors exceeding 30 metres in length in shop premises, or 45 metres in office and factory premises, should be sub-divided by **fire doors** to prevent the free travel of smoke and other products of combustion throughout the length of the corridor (see diagram 6 and Table A). The sub-division should be such that, wherever possible, no undivided length of corridor is common to more than one **storey exit.** Doors provided for the sole purpose of restricting the passage of smoke need not be **fire doors** provided that they are fitted with suitable smoke seals, are of substantial construction and self-closing and double swing as necessary. Care should be taken to ensure that smoke cannot readily by-pass these doors eg above a false ceiling or via alternative doors from a room or doors from adjoining rooms opening on both sides of the corridor.

*Note: Generally, false ceilings should be provided with **cavity barriers** or **smoke stopping** barriers over any **fire doors**. However, where the false ceiling itself is an integral part of the fire-resisting construction, this may not be necessary.*

Diagram 6   Example of the sub-division of a corridor by **fire doors**

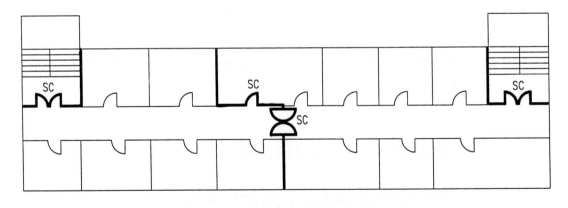

───── Fire-resisting walls & fire doors.

SC   Self-closing.

*Note: Where the sub-dividing doors are not **fire doors** the lines shown as fire-resisting in the diagram should be at least the same fire resistance as the doors.*

14.32 The floor of any corridor or passage should not be inclined at a gradient steeper than 1 in 12 to the horizontal.

Travel within stairways and to final exits

14.33 Any stairway, lobby or corridor which forms part of the **means of escape** from the premises, should be constructed and arranged so as not to impede the free flow of persons using the route. The escape route should also be of a uniform and adequate width for the maximum number of persons likely to use it. The following considerations regarding exit widths also apply:-

(a) ideally, stairways should be at least 1.05 metres wide; and

(b) where wheelchair users may be present, exits should be not less than 900mm wide.

14.34 Existing stairways wider than 2.1 metres should normally be divided into sections, each separated from the adjacent section by a handrail, so that each section measured between the handrails is about 1.05 metres wide.

14.35 Stairways and landings should be measured, with no account being taken of skirtings, as follows:-

(a) when enclosed on each side by walls, from the finished surface of the wall on one side to the finished surface of the wall on the other side;

(b) when constructed with a wall on one side only, from the finished surface of the wall to the outer edge of the steps and landings; and

(c) when provided with balustrades on both sides, from the outer edge of the steps or landings on one side to the outer edge of the steps or landings on the other side.

Side handrails projecting nor more than 100mm and central handrails of not more than 150mm in total width may be disregarded for calculation purposes.

Number of stairways

Buildings with a single stairway

14.36 It will normally be necessary for a building to be provided with two or more stairways, but a single stairway may be considered satisfactory in the circumstances described in paragraphs 14.37 and 14.38.

Shops and offices

Shops and offices which are of normal fire risk in which no floor has an area in excess of 90m², and if divided or partitioned, the exit from each floor is clearly visible from any part of the floor area and the building contains not more than two floors, one of which may be a basement

14.37 (a) In these circumstances, the stairway may be an open stairway provided that:-

(i) the stairway discharges not more than 3 metres from a **final exit**;

(ii) the building contains no public restaurant or licensed bar;

(iii) the **distance of travel** from any point in the building to the **final exit** does not exceed that shown in line Ei of Table D; and

(iv) in the case of a stairway serving any basement floor, the floor is not more than 3.5 metres below the ground floor level (see diagram 7).

Diagram 7 Example of shop or office of two floors with no floor area in excess of 90m².

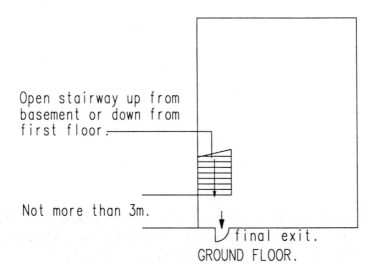

Shops and offices which are of normal fire risk in which no floor has an area in excess of 90m², and if divided or partitioned, the exit from each floor is clearly visible from any part of the floor area and the building contains three floors, one of which is a basement

(b) In these circumstances, one part of the stairway may be an open stairway provided that:-

(i) the stairway which serves either the basement or first floor is enclosed with **fire-resisting** construction at the ground floor; or

(ii) the enclosed stairway discharges to a **final exit** which is independent of, and separated from, the ground floor accommodation; or

(iii) the recommendations in (ii) and (iii) of paragraph 14.37(a) apply (see diagram 8).

Diagram 8 Example of shop or office of three floors with no floor area in excess of 90m².

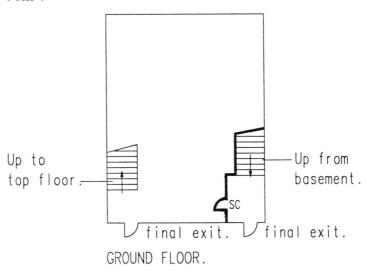

Shops or offices which are of normal fire risk in which the area of any floor exceeds 90m² but does not exceed 280m² and which consists of not more than a basement, a ground floor and a first floor

(c) In these circumstances, the stairway should be a **protected stairway** discharging via a **final exit** independent of, and separate from, the ground floor accommodation (see diagram 9).

Diagram 9   Example of shop or office of three floors with no floor area in excess of 280m².

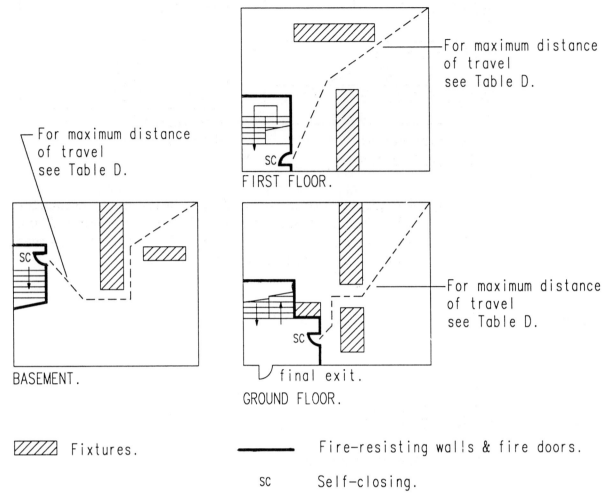

Shops and offices which are of normal fire risk and have floor areas in excess of 280m² with not more than 4 floors above the ground floor

(d) In these circumstances, the stairway will need to be separated from the remainder of the building in the manner recommended in paragraphs 14.39 and 14.44 (see also diagram 12 and paragraph 14.46). Where it is impracticable to achieve two **fire door** separation in accordance with paragraph 14.39 in a building with not more than 2 floors above the ground floor, the stairway should be made a **protected route** and suitable automatic fire detection arrangements provided in the building (see paragraphs 15.12 to 15.14).

Factories

14.38 (a) A single stairway may be satisfactory in a factory of low fire risk if there are not more than two floors above the ground floor (see diagram 10). If a basement is provided paragraphs 14.51 to 14.54 should be consulted.

Diagram 10 Example of low fire risk factory

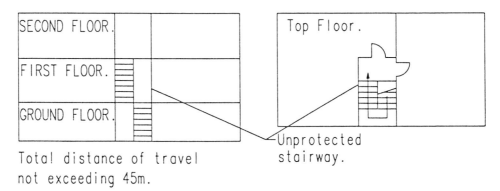

SECOND FLOOR.
FIRST FLOOR.
GROUND FLOOR.
Top Floor.
Unprotected stairway.
Total distance of travel not exceeding 45m.

(b) A single stairway may be satisfactory in a factory of normal fire risk if there is not more than one floor above the ground floor (see paragraph 14.46 and diagram 11). If a basement is provided paragraphs 14.51 to 14.54 should be consulted.

Diagram 11 Example of normal fire risk factory.

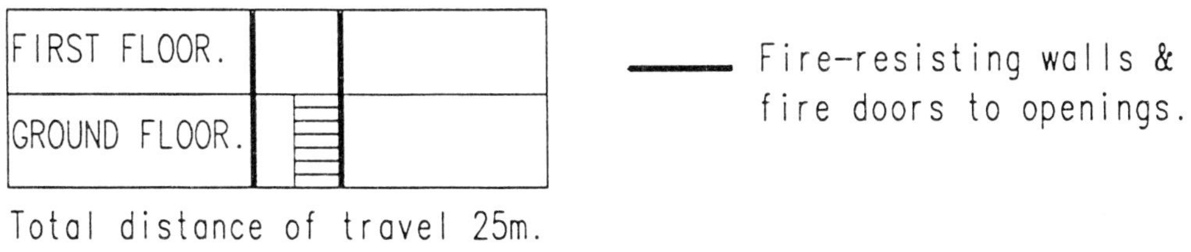

FIRST FLOOR.
GROUND FLOOR.
Total distance of travel 25m.
Fire-resisting walls & fire doors to openings.

Protected lobby approach to single stairway building

14.39 With the exception of certain small buildings (see paragraphs 14.36 to 14.38), in a building with a single stairway the access to the stairway from any floor, (other than the top floor) or any room (other than a toilet containing no fire risk) should be through a **protected lobby** or a **protected corridor** (see diagram 12).

Diagram 12 — Example of single stairways separated from floor areas by (a) a **protected lobby** or (b) a **protected corridor**.

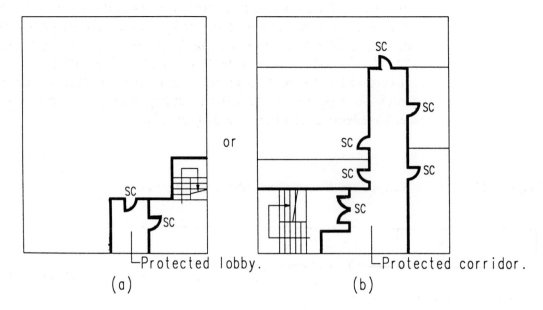

— Fire-resisting walls & fire doors.
SC Self-closing.

Buildings with two or more stairways

14.40 There should be two or more stairways for the **means of escape** in all buildings which do not meet the criteria of paragraphs 14.37 and 14.38.

Width of stairways

14.41 In normal circumstances, stairways should be not less than 750mm wide and in all cases the aggregate capacity of stairways should be sufficient for the number of persons likely to have to use them at the time of a fire. In this connection it will be necessary to consider the possibility of one stairway being inaccessible because of the fire and the aggregate capacity should allow for this possible reduction (see also paragraph 14.33).

Enclosure of stairways

All buildings

14.42 Other than the situations described in the relevant parts of paragraphs 14.37, 14.38 and 14.48 (a), all escape stairways should be separated from the remainder of the building by **fire-resisting** construction and **fire doors** so as to form a **protected stairway.**

14.43 The method whereby a stairway is separated from the remainder of the building should be such as to ensure that a person need not pass through a stairway enclosure to reach an alternative escape route. If this is not possible then the stairway should still be separated and it may be reasonable for an alternative route to by-pass the stairway by means of balconies (see paragraph 14.78) or by means of a by-pass corridor or exceptionally intercommunicating doors between rooms (see paragraph 14.77). By-pass corridors and doors should be of appropriate **fire resistance** (where necessary) and of suitable width (see paragraphs 14.22 to 14.24). By-pass or intercommunicating doors should be unobstructed and available at all times (see diagram 13).

Diagram 13   Example of by-pass route around a stairway.

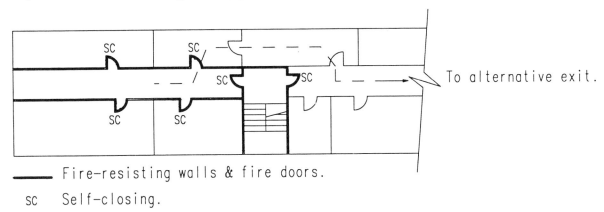

14.44 Ideally stairway enclosures should lead direct to a **final exit**. Where there is only one stairway from the upper floor(s) of a building and a **final exit** cannot be provided from the stairway enclosure, one of the following arrangements should be adopted:-

(a) the provision of 2 exits from the stairway enclosure each giving access to **final exits** by way of routes separated from each other by **fire-resisting** construction (see diagram 14); or

(b) the provision of a **protected route** from the foot of the stairway enclosure leading to a **final exit** (see diagram 15).

Diagram 14 Example of separate routes from stairway enclosure to separate **final exits**.

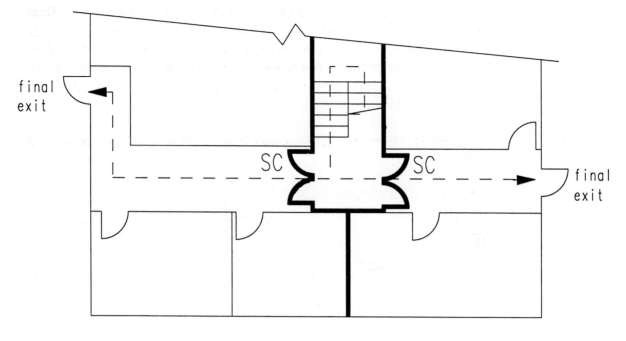

— Fire-resisting walls & fire doors.

SC Self-closing.

Diagram 15 Example of **protected route** from stairway enclosure to a **final exit**.

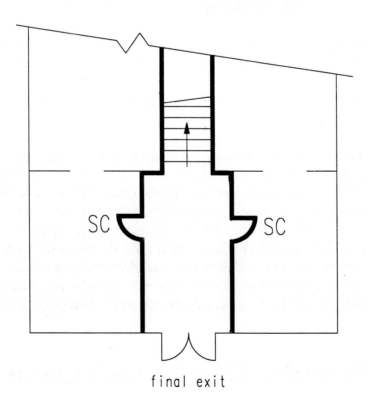

— Fire-resisting walls & fire doors.

SC Self-closing.

14.45 Where from an upper floor(s) of a building there is more than one stairway, which is required by paragraphs 14.48 and 14.49 to be separated from the remainder of the building, and the stairways do not have **final exits** from the stairway enclosures, the stairways and the routes to their respective **final exits** should be separated from one another by **fire-resisting** construction and **fire doors** in such a way that an outbreak of fire at any point cannot affect more than one escape route from the stairways (see diagram 16).

Diagram 16 Example of alternative escape routes from stairway enclosures separated by **fire-resisting** construction.

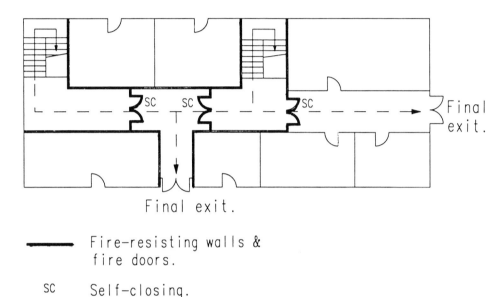

―――― Fire-resisting walls & fire doors.

SC Self-closing.

14.46 Where a stairway is a **protected route** it will not be necessary to have regard to the **distance of travel** beyond the **storey exit**. Where a stairway is not a **protected route**, it will be necessary to include the stairway in the total permitted **distance of travel** (see lines Ei and Eii of Table D).

14.47 In a building which is in multi-purpose use, where stairways are common to all occupancies, the standard of fire protection of the stairways will be determined by the occupancy which requires the higher standard. For example, in a building in which there are four floors above the ground floor and there is a factory occupying the basement, ground and first floor with offices above, the stairways would need to conform with the recommendations contained in paragraph 14.48 regardless of the fact that the factory premises do not exceed more than one floor above the ground floor specifically mentioned in paragraph 14.48(a)(ii).

Normal and high fire risk buildings

14.48 (a) With the exception of stairway arrangements in buildings described in paragraphs 14.37 and 14.38 all other stairways in buildings of normal or high fire risk should be separated from the remainder of the building by **fire-resisting** construction and by **fire doors** unless:-

    (i) the stairway is an **accommodation stairway** (see paragraph 14.56); or

    (ii) in the case of a factory, the stairway forms an alternative escape route which is not the only escape route from the building (or part of the building), and the building has not more than one floor above the ground floor.

(b) A stairway so separated from the remainder of the building should be regarded as a **protected route** if it discharges via a **final exit** and the only doors into it are:-

    (i) from toilets containing no fire risk (see Note 3 to Table A);

    (ii) from **protected lobbies**;

    (iii) from corridors; or

    (iv) from lift wells with no openings other than those to the stairway enclosure.

(c) Exceptionally, however, doors other than those referred to in (b) above, may prove acceptable if, for example, they are doors from rooms of low fire risk that open into the stairway. If there is no **final exit** from the stairway it may still be possible to regard it as a **protected route** if it conforms to the recommendations in paragraph 14.44 in the manner illustrated in diagrams 14 and 15.

Factory of low fire risk

14.49 In a factory of low fire risk, any stairway which serves more than two floors above the ground floor should be separated from the remainder of the building by **fire-resisting** construction and by **fire doors**. The stairway may be regarded as a **protected route** if there is a **final exit** from it.

High rise buildings

14.50 In any factory building having any floor more than 18 metres above ground level or in any office or shop building having any floor more than 24 metres above ground level the only doors into a stairway should be:-

(a) from toilets containing no fire risk – see Note 3 to Table A;

(b) from **protected lobbies**; or

(c) from lift wells contained within a stairway enclosure.

A stairway separated in this way (**protected lobby** approach) and having a **final exit** from the stairway should be regarded as a **protected route** (see diagrams 12 (a) and (b)).

Diagram 17   Example of a stairway separated from the floor area by a **protected lobby**.

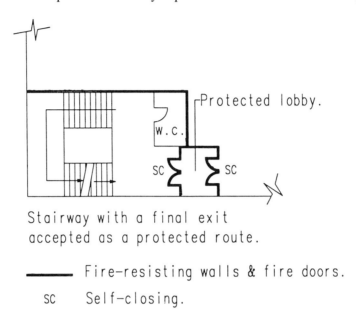

Stairway with a final exit accepted as a protected route.

—— Fire-resisting walls & fire doors.
SC   Self-closing.

Stairways from basements.

14.51 Basements require special consideration as their stairways provide an upward escape route which may be filled with smoke.

14.52 In all buildings, other than those dealt with in paragraphs 14.37 and 14.38, it is preferable that a stairway serving upper floors should not extend to the basement. Wherever possible all stairways to basements should be entered at ground level from open air, and in such positions that smoke from any fire would not obstruct any exit serving the upper floors of the building. Where the stairway links a basement with the ground floor the basement should be separated from the ground floor preferably by two 30 minute **fire doors**, one at basement and one at ground floor level.

14.53 Exceptionally in a factory, the basement may be separated by one 60 minute **fire door** at ground level, or where the basement is small and does not present a high fire risk, it may be acceptable to provide one 30 minute **fire door**.

14.54 In low and normal fire risk premises, a single stairway from a basement may be satisfactory provided that the **distance of travel** from any point in the basement to a **final exit** is within the acceptable limits (see Table D).

14.55 In high fire risk premises there should be an alternative stairway from a basement to ground level, unless there is a suitable alternative route to a **final exit** from the basement.

Diagram 18 Example of basement separated from the ground floor by 2 x 30 minutes **fire doors**, one at the foot of the stairway and one at its head.

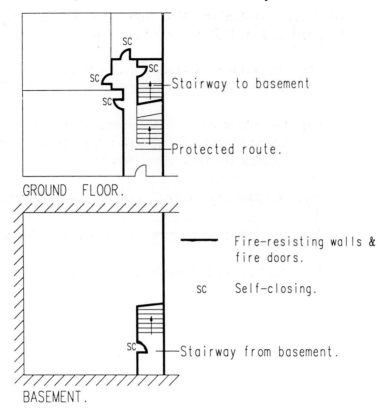

Diagram 19 Example of basement separated from the ground floor by 2 x 30 minutes **fire doors**, one between any room in the basement and one at the head of the stairway at ground floor.

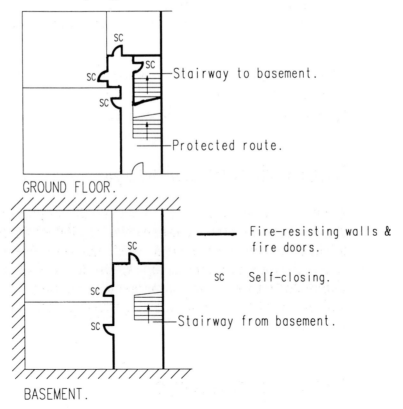

Accommodation stairways

14.56 Accommodation stairways need not be separated from the remainder of the building provided that:-

(a) the **means of escape** are sufficient without relying on the **accommodation stairway**;

(b) no escape route from a **dead-end** of an upper floor passes the access to an **accommodation stairway** (see also paragraph 12.7 in relation to other floor openings);

(c) the stairways do not pass from one **compartment** to another;

(d) no such stairway serves more than two floors; and

(e) neither of the floors is connected by an open stairway to a third floor.

Escalators and travolators

14.57 Escalators and travolators are not normally acceptable as a **means of escape**. However, they may need to be taken into consideration in certain occupancies such as railway premises or where they are enclosed within a protected shaft.

14.58 Escalators and travolators not within stairway enclosures may need to be separated from the remainder of the building by **fire-resisting** construction and by **fire doors**. However, in certain premises eg shops, it may be reasonable to regard them as similar to **accommodation stairways** (see paragraph 14.56).

14.59 To avoid a situation where persons are carried towards a fire it is important that arrangements should be made to stop escalators and travolators at the outset of an emergency.

Ventilation of stairways

14.60 Wherever practicable, there should be provision for ventilating stairways in the event of a fire, particularly if the stairway enclosure is not adjacent to an external wall which has openable windows, and the stairway continues uninterrupted to the top of the building. The minimum area of permanent or openable venting should not be less than 1m^2 or 5% of the plan area of the stairway enclosure, whichever is the greater.

External stairways

14.61 Where an external escape stairway is provided it should be a **protected route** and it will be necessary to ensure that the use of it at the time of a fire cannot be prejudiced by smoke and flames issuing from openings (eg windows and doors) in the external wall of the building below and adjacent to the stairway. Any door opening onto the stairway below the top floor and any door in the external wall beneath the stairway should have a minimum **fire resistance** of 30 minutes and be self-closing. In situations where windows are less than 1.8 metres horizontally from the stairway, they should be of a fixed type and have a minimum **fire resistance** of 30 minutes. The **fire resistance** should apply to the entire wall including doors and windows. It will also be necessary to provide lighting and consider the protection of the stairway from the weather (see diagram 20). Exceptionally it may be possible to accept a limited number of unprotected windows provided that the rooms in question are separated from the remainder of the building by **fire-resisting** construction so as to afford adequate separation between internal and external stairways.

Diagram 20 Example of defined zone for **fire-resisting** windows and doors.

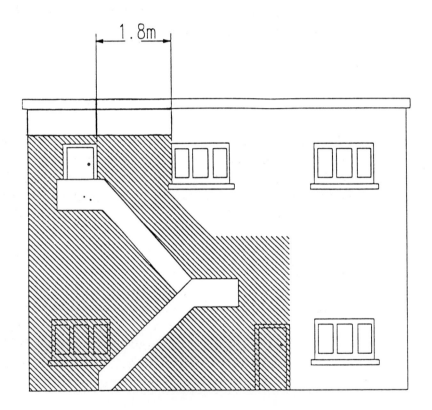

Headroom of stairways, lobbies, corridors, passages

14.62 In general all **means of escape** should have a clear headroom of not less than 2 metres. There should be no projections from any wall or ceiling below this height (except for door frames) which could impede the free flow of persons using the escape route.

Handrails

14.63 A continuous handrail should be fixed on each side of all stairways, steps, landings and ramps at a height suitable for the average person eg about 840mm to 1 metre above the pitch line of the nosings of steps. Where such stairs, ramps or flights of steps are wider than 1 metre a second handrail should normally be provided. The handrail should not normally project more than 100mm and the ends should be turned for safety (see paragraph 14.35).

Spiral or helical stairways

14.64 Spiral and helical stairways may form part of the **means of escape** provided that they are designed in accordance with British Standard 5395: Part 2 and, if they are intended to serve members of the public, should be a type E (public) stair, in accordance with that standard (but see paragraph 14.66).

14.65 The going of a step should be not less than 250mm and all consecutive tapered treads should have:-

(a) the same minimum going of 250mm, measured at a point 270mm from the wide end of the tread (see Approved Document K to the Building Regulations 1991 for new stairways or in Scotland Part S of the Technical Standards associated with the Building Standards (Scotland) Regulations);

(b) the same rate of going and taper;

(c) their narrow end at the same side of the flight; and

(d) an angle of taper of not more than 15 degrees.

14.66 Existing spiral stairways which do not conform to British Standard 5395 Part 2 are acceptable only in exceptional situations eg for use by not more than 50 able-bodied adults who are not members of the public. The stairway should be not more than 9 metres in height, nor less than 1.5 metres in diameter and should give adequate headroom. The handrail should be continuous throughout the full length of the stairway.

Ladder devices

14.67 Portable ladders and throw out type ladders are not suitable as **means of escape** for members of the public. However, there may be some work situations in which it will be reasonable for ladders of this kind to provide escape for one or two able-bodied and agile persons eg from an item of high level plant, from an overhead travelling crane or a hoist in a factory.

14.68 Fixed vertical or raking ladders should not be used by members of the public. They are only acceptable for use by limited numbers of persons who are able-bodied and active enough to be able to use them without difficulty. Where such ladders form the means for gaining access to plant or machinery rooms, it will be reasonable to accept them for **means of escape** purposes for persons who are expected to use them in normal circumstances. Where vertical ladders are accepted, they should be suitably guarded and the total descent by such means should not generally exceed 9 metres without an intermediate landing.

Lowering lines and other self rescue devices

14.69 Lowering lines are not suitable for use by members of the public. Automatic lowering lines and other manipulative emergency devices for self-rescue should be contemplated only in very exceptional circumstances, for example, to afford some alternative emergency arrangement in a factory for one able-bodied person who could be trained in the use of the device and who is required to work in an isolated position where it would be unreasonable to look for conventional routes of escape.

Escape from tall items of plant

14.70 Reasonable **means of escape** should be provided from any tall items of plant within a factory building. Plant of this kind will normally entail employees using such facilities as vertical and raking ladders and possibly in some cases even less conventional means of gaining egress. Where this is the case, it will be satisfactory to rely on such arrangements as an alternative escape route (see paragraphs 14.67 to 14.69).

Ramps

14.71 Where a ramp forms part of an escape route it should have an easy gradient and in no case should it be steeper than 1 in 12. Handrails and non-slip surfaces should be provided to guard against a person slipping. This will be particularly necessary where a ramp is exposed to the weather. Guard-rails should be provided where the ramp is more than 600mm above the ground.

Lifts and hoists

14.72 Lifts and hoists are not normally acceptable for **means of escape**. Exceptionally, however, when it is necessary to consider **means of escape** for people with disabilities, the use of a lift(s) should be considered provided that the appropriate provisions of British Standard 5588: Part 8 are complied with.

14.73 Unless a lift is situated within a stairway enclosure which is a **protected route** it should be contained within a lift well enclosure of **fire-resisting** construction in which the access doors are **fire doors**. Existing sliding doors to lift shaft openings are sometimes ill fitting in their slides and frames and offer a poor barrier to the passage of smoke. In such cases where the opening discharges into a corridor which is a **dead-end**, a **fire-resisting** screen and **fire doors** should be provided at the entrance to the lift. A person should not have to pass through the lobby so formed to reach the continuing route of escape (see diagram 21).

Diagram 21 Example of **fire-resisting** protection to lift entrance discharging into **dead-end** corridor

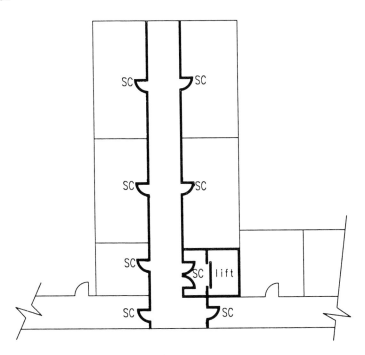

─── Fire-resisting walls & fire doors.
SC   Self-closing.

14.74 A lift motor situated at the foot of a lift well which is within the enclosure of a **protected stairway** which forms the only escape route from a building (or part of a building) should be housed in a **compartment** separated from the lift well by **fire-resisting** construction. Any opening necessary in the separation between the **compartment** and the well for the operation of the lift should be as small as possible.

14.75 Where practicable, a lift well should have a permanent ventilation at the top equal to not less than $0.1m^2$ for each lift in the lift well enclosure.

Wall and floor hatches

14.76 Only in exceptional circumstances should it be necessary to rely for **means of escape** on wall hatches and floor hatches. However, there may be some instances when, because of structural difficulties, it will be reasonable to accept arrangements of this kind for a very limited number of persons who are active enough to use them, but under no circumstances should they be provided for the use of members of the public. Where wall and floor hatches are provided, there may be a need to take special precautions to safeguard against their obstruction and protect their use eg by the provision of guard rails round the hatchway.

Intercommunication between rooms, occupancies and buildings

14.77 It may be necessary to provide **means of escape** by way of intercommunication between adjoining rooms, occupancies or buildings. Normally this should be done by providing a doorway of minimum width appropriate to the circumstances (see paragraph 14.22) and also having, where necessary, a door of the requisite degree of **fire resistance** (see paragraph 12.1). Where it could be difficult or it might not be reasonable to provide a doorway of normal height and width, it may be possible to accept an escape hatch or similar unconventional exit (see paragraph 14.76). Where intercommunication in this way between different occupancies is being considered it will be necessary to check that a legally binding agreement is in force between all interested parties so that the use of the route will be available at all times. This is particularly important if the different occupancies use their premises at different times.

Balconies, bridges and walkways

14.78 A balcony, bridge or walkway can sometimes be used to by-pass a stairway enclosure. Where they form any part of the **means of escape** it will be necessary to ensure that their use at the time of a fire cannot be prejudiced by smoke or flames issuing from openings (eg windows and doors) in the external wall of the building. It is essential that a door to or from a balcony, bridge or walkway which is intended for use as a **means of escape** is kept unlocked. In all cases there is a need to ensure that these **means of escape** can be used safely by providing, as appropriate, guard rails, hand rails, toe boards, etc. Depending on the circumstances it may be necessary to provide weather protection and **emergency escape lighting** (see also paragraphs 14.101 and 14.102).

Roof exits and exits at upper levels

14.79 Other than from a basement, persons escaping from a fire, should not normally have to ascend to a higher level to reach a **place of safety**. Exceptionally, where such a route is necessary, **final exits** at an upper level or roof exits may be acceptable provided that the following criteria are met:-

Roof exits

(a) roof exits (other than across flat roofs to a properly constructed stairway) should not normally be used by members of the public;

(b) the escape route should normally be flat. It should be adequately defined, lit (see paragraphs 14.101 and 14.102) and the surface should be of a safe non-slip character and the route guarded with protective barriers;

(c) the escape route across the roof and its supporting structure should be constructed as a **fire-resisting** floor;

(d) if the escape route is in one direction only, any ventilation outlet or other extract system and any door roof lights, or windows that are not **fire-resisting** should not be sited within 3 metres of the route; and

(e) the exit from the roof should be in, or should lead directly to a **place of safety.**

Note: See paragraph 14.77 about the need to ensure that there is a legally binding agreement on the use of the route.

Upward escape

(f) other than from a basement, persons should not normally have to ascend more than one level (ie from one floor to the next or from a top floor to roof level);

(g) the higher level to which persons ascend should not be to an area of high fire risk;

(h) the route by which the higher level is gained should be a stairway; and

(i) if the stairway also serves any lower level(s), the upward escape route should be separated from the remainder of the stairway by **fire- resisting** construction and by **fire doors.**

Wicket doors and gates

14.80 Often a wicket door is provided in a large door or in a shutter. These may be satisfactory for up to 3 persons for the purpose of **means of escape** in high fire risk buildings (or parts of buildings). Elsewhere they may also be acceptable provided that the numbers of persons likely to use them are not more than 15, and are not members of the public. Unless circumstances are exceptional, a wicket door should provide a minimum opening of which the top is not less than 1.5 metres and the bottom of which is not more than 250mm above the floor. The width of the opening should preferably be not less than 500mm and in no case less than 450mm.

Goods delivery doors, shutters etc

14.81 Loading and goods delivery doors, shutters (roll, folding or sliding), up-and-over doors and similarly filled openings do not normally provide satisfactory exits for **means of escape** in case of fire. However, there may be instances in buildings (or parts of buildings) of low or normal fire risk where it will be possible to regard them as such provided that they are not likely to be obstructed and can be opened manually even if normally power operated. Where these doors are used as a **means of escape** they should be capable of being easily and immediately opened by persons escaping (see also paragraph 14.86).

Window exits

14.82 Only in exceptional circumstances should windows be used as a **means of escape** to a **place of safety.** Where they are provided (see diagram 22), the following conditions should be met:-

(a) the exits should not be used by members of the public;

(b) the exit should not be used by more than 10 able-bodied people;

(c) any such window should be of the casement type sufficiently large and openable to permit any persons to pass through without undue difficulty (eg at least 850mm x 500mm wide in the clear with the casement open);

(d) suitable steps should be provided up to the windowsill, both inside and outside the building, with hand grips provided as necessary;

(e) the external surface should be level and unobstructed and the route of escape should lead to a **place of safety**;

(f) the height of the windowsill should be not more than 1.1 metres above the floor level of the room it serves and of the ground or external level surface upon which it discharges;

(g) where the window forms part of a roof exit, the provisions of paragraph 14.79 should be applied;

(h) simple fastenings which do not require the use of a key in an emergency should be fitted to the openable window; and

(i) any such window should be conspicuously indicated as a "Fire Exit" (see paragraph 14.94).

Diagram 22: Example of a window escape

Doors on escape routes and their fastenings

14.83 (a) Wherever possible, a door provided for **means of escape** should open in the direction of travel. It should always do so if:-

(i) it is from a room in which a fire may develop very rapidly; or

(ii) the door is from an area from which more than 50 persons may be required to escape.

(b) The door should also:-

(i) be hung so that, when open, it does not obstruct any escape route;

(ii) open through not less than 90 degrees;

(iii) be provided with a vision panel if it is hung to swing both ways; and

(iv) if it is a **fire door**, protecting an escape route, be fitted with smoke seals.

Self-closing devices for fire doors

14.84 All **fire doors** except those to cupboards and service ducts, should be fitted with effective self-closing devices to ensure the positive closure of the door. Rising butt hinges are not normally acceptable. **Fire doors** to cupboards, service ducts and any vertical shafts linking floors should be either self-closing or should be kept locked shut when not in use and labelled accordingly (see paragraphs 14.97 and 14.98).

Automatic door releases

14.85 It is acknowledged that in some existing buildings, self-closing **fire doors** may cause difficulties for staff and members of the public. In such circumstances, automatic door releases can be used to hold open such doors provided that:-

(a) the door release mechanism conforms to British Standard 5839: Part 3 and fails safe (ie in the event of a fault or loss of power the release mechanism is triggered automatically);

(b) all doors fitted with automatic door releases are linked to an automatic fire warning system appropriate to the fire risk in the premises and complying with British Standard 5839: Part 1 (see Note 1);

(c) all releases are automatically triggered by any one of the following:-

(i) the actuation of any automatic fire detector;

(ii) the actuation of any manual fire alarm call point;

(iii) any fault which renders the fire warning system inoperable; or

(iv) the isolation of the alarm system for any reason eg maintenance.

(d) doors so fitted are capable of being closed manually;

(e) the release mechanisms are tested at least once each week in conjunction with the fire alarm test to ensure:-

(i) that the mechanisms are working effectively; and

(ii) the doors close effectively onto their frames; and

(f) the automatic door release is fitted as close as possible to the self-closing device in order to reduce the possibility of the door becoming distorted.

Notes: 1. In exceptional circumstances and where the number of doors involved is very small, it may be reasonable to accept a more localised method of operation eg smoke detectors on each side of the door opening provided they are linked to the fire alarm system.

2. It is essential that doors fitted with such devices are maintained free from obstruction.

3. All doors should be closed at night.

4. The fire warning system and the automatic door release mechanisms should be regularly maintained by a competent person.

5. If there are staff with disabilities it may be considered that door closing devices should be of the delayed action type.

Fastenings on doors

14.86 Doors used for **means of escape** should be kept unlocked at all times when persons are in the building and in no case should a door be so fastened that it cannot be easily and immediately opened by persons escaping without the use of a key.

14.87 Wherever possible there should only be one fastening. However, where more than one fastening cannot be avoided, all but the single emergency fastening should be kept released at all times persons are on the premises (see paragraph 14.89).

14.88 Where the door:-

(a) may be used at the time of a fire by more than 50 persons; or

(b) is an exit from an area in which fire may develop very rapidly

and has to be kept fastened while persons are in the building, it should be fastened only by pressure release devices such as panic latches, panic bolts or pressure pads which ensure that the door can be easily and immediately opened by persons within. Panic bolts and panic latches should comply with British Standard 5725: Part 1.

14.89 Where in cases other than those described in paragraphs 14.86 and 14.87 there is a need to fasten doors for security reasons, some other form of fastening may be acceptable. Any such fastenings will need to ensure that the door can be easily opened by persons escaping. It will also be necessary for persons who may have to use the door to understand the method of operation. Normally this type of fastening is not suitable for areas accessed by unaccompanied members of the public.

Note: Where a special form of mechanical fastening is used, a notice providing clear instructions as to its use should be prominently displayed. The use of electrically operated fastenings requires special consideration to ensure that they are readily operable at all material times and will fail safe in the open position.

Revolving doors

14.90 A revolving door should not be provided specifically as an emergency exit. Such a door may only be considered as a **means of escape** if it automatically converts to outward opening upon pressure from within.

14.91 Where a revolving door, not complying with paragraph 14.90, is installed it should be supplemented by additional, and immediately adjacent, exit doors which are clearly indicated as such. Exceptionally, however, where only a small number of persons are likely to be involved, it may be possible for exit doors which are not immediately adjacent to be accepted.

Sliding doors

14.92 A sliding door should not normally be provided specifically as an emergency exit (but see Note below). Manually operated sliding doors are only acceptable on an escape route for use by small numbers of employees or accompanied members of the public. Where power operated sliding doors are provided they should be capable of either converting to outward opening doors under reasonable pressure or should be provided with a monitoring system to ensure that they fail safe to the fully open position in the event of either the failure of the power supply or the actuation of the fire warning system.

Note: Where the use of a sliding door has been previously approved by a fire authority, the use of such a door continues to be acceptable unless there has been a material change to the premises.

Exit and directional signs

14.93 All fire safety signs and graphic symbols should comply with British Standard 5499: Parts 1 and 3. However, existing signs need not be replaced immediately if they are fulfilling their purpose effectively. They should, however, be examined and replaced if they are found to be inadequate.

14.94 Any exit which is not a normal route of travel from a building should be clearly indicated. This can take the form of a pictogram or a sign bearing the words "Fire exit" in conspicuous lettering. The symbol or sign should, wherever possible, be displayed immediately above the exit opening. Where this is not possible a position should be chosen where the symbol or sign can be clearly seen and it is least likely to be obstructed.

Note: There could be circumstances where the provision of photoluminescent signs, self luminous signs, or other wayfinding systems could be advantageous, to supplement conventional signs.

14.95 Where an exit cannot be seen or where a person escaping might be in doubt as to the location of an exit eg a shop where display stands and other obstructions may prevent a clear view of the exit doors, "Fire exit" signs, to include a directional arrow, should be provided at suitable points along an escape route. Such signs should be sufficiently large, should be fixed in conspicuous positions, and wherever possible be positioned between 2 metres and 2.5 metres above the floor level.

Signs on doors

14.96 A sign with the words "Push bar to open" should be permanently displayed either on or immediately above the push-bar on all doors fitted with a panic bolt or panic latch. Other devices should have a suitable sign describing the method of operation.

14.97 A sign with the words "Fire door–Keep shut" should be permanently displayed at about eye level on both faces of all **fire doors** except those to cupboards. **Fire doors** which are normally held open but which close automatically on the operation of fire detectors should bear the words "Automatic fire door–Keep clear".

14.98 A sign with the words "Fire door–Keep locked shut when not in use" should be permanently displayed on the outside face of all **fire doors** not required to be self closing eg cupboards.

14.99 A sign with the words "Fire escape–Keep clear" should be permanently displayed at about eye level on the external face of all doors which are provided solely as a **means of escape** in case of fire and which, because they are not normally used, could be obstructed.

Note: All fire signs should be in conspicuous lettering and of appropriate size.

Lighting of escape routes

14.100 All escape routes from the building, including external ones where appropriate, should be provided with sufficient artificial lighting for occupants to see their way out safely when there is not enough natural light. Control switches should be provided at the entry point of an escape route and clearly indicated when not provided for normal use.

Emergency escape lighting

14.101 **Emergency escape lighting** should be provided in those parts of buildings where there is underground or windowless accommodation, core stairways or extensive internal corridors. Generally the need for **emergency escape lighting** will arise more frequently in shops than in factories and offices because of the greater likelihood of people in the building being unfamiliar with the **means of escape**. Any new **emergency escape lighting** system should conform with the appropriate provisions including the certification, testing and servicing provisions of British Standard 5266: Part 1.

14.102 Notwithstanding the recommendations at sub clause 9.3.2 of British Standard 5266: Part 1, an existing **emergency escape lighting** system may be acceptable if it is automatic in operation on the failure of the mains sub-circuit supply and is capable of maintaining the required level of illumination for two hours, although acceptance of not less than one hour should not be discounted in small buildings ie those having not more than two floors above the ground floor and where the escape routes are straightforward.

Ventilation systems

14.103 Where ventilation systems may assist the spread of fire, smoke or hot gases, it will be necessary to take steps to safeguard the **means of escape** (see paragraph 12.7).

Notes/Amendments

Chapter 15: **MEANS FOR GIVING WARNING IN CASE OF FIRE AND FOR DETECTING FIRE**

General

15.1 Any building containing parts to which this guide applies should be provided with a means for giving warning in case of fire to those persons who are in those parts of the building.

Fire alarm systems

15.2 A fire alarm system should normally comply with the recommendations for Type M of British Standard 5839: Part 1 and where such an alarm system is installed an installation and commissioning certificate should be provided. However, there may be situations where a simpler means of raising the alarm could be regarded as adequate.

Manually operated alarms

15.3 Manually operated sounders eg rotary gongs, may be acceptable in small buildings if each sounder can be sited in a safe position and will give a warning which is audible throughout the building.

Staged fire alarms

15.4 In the main, fire alarm systems are single-stage, ie the system requires the simultaneous evacuation of all of the occupants of a building. However, in certain premises, a "staged" alarm system may be more appropriate. The options available are either a two-stage evacuation or a phased evacuation.

Two-stage evacuation

15.5 Two-stage fire alarm systems allow for staged evacuation ie evacuation of persons in the area of immediate risk from fire and total evacuation only when it becomes necessary. The first stage of a two-stage system gives an "evacuation" signal which is continuous, in or near the zone of origin. At the same time, other parts of the building receive an alert signal (usually intermittent), which should not be treated as a signal for general evacuation. The intermittent signal can be switched to a continuous evacuation signal which requires the total simultaneous evacuation of the remainder of the building (the second stage). The approval of the fire authority should be obtained before such a system is installed and any system should comply with British Standard 5839: Part 1.

Phased evacuation

15.6 A phased evacuation system is an arrangement for the continuous "evacuation" signal to be sounded on the floor of origin and the floor directly above. Other floors are made aware of a fire by an alert signal and a public address message. Subsequently, and if necessary, the evacuation signal will be sounded on other floors, usually above the floor of origin and progressive evacuation will follow, floor by floor. An exception is a basement area where, if the basement is in the zone of origin, all floors below ground floor level should normally be evacuated. Since schemes using phased evacuation may involve a reduction in escape widths they should not normally be considered without the provision of adequate compartmentation, sprinklers, floor marshals and a fire control centre. The approval of the fire authority should be obtained before such a system is installed and any system should comply with British Standard 5839: Part 1.

Staff alarms

15.7 In certain premises eg large shops, an initial general alarm may be undesirable because of the number of public present and the need for fully trained staff to effect pre-planned procedures for safe evacuation. Actuation of the fire alarm system will therefore cause staff to be alerted eg by discreet sounders, personal paging systems etc. Provision will normally be made for full evacuation of the premises by sounders or a message broadcast over the public address system. In all other respects, any staff alarm system should comply with British Standard 5839: Part 1.

Call points

15.8 Call points for electrical alarm systems should comply with British Standard 5839: Part 2, and these should be installed in accordance with British Standard 5839: Part 1.

Fire warning signals

15.9 The evacuation signal incorporated into the electrical fire warning system should be distinctive and the sounders etc. should be sited so as to ensure that a common warning signal is perceptible throughout the building or throughout all parts in which there is a requirement to provide such a warning system. In premises where the noise level may be excessive, or in any other situation where a normal type of sounder may be ineffective, visual signals may be used to supplement the audible alarms. They should not, however, be used on their own.

15.10 In larger premises, where a public address system is provided, the fire alarm system should normally be supplemented by verbal instructions. Evacuation messages should be brief, positive and clear.

Audible alarms by intercommunication or public address equipment

15.11 Where public address equipment is used to transmit a general alarm, the warning signal should take priority, override other facilities of the equipment and be distinct from other signals which may be in general use. The alarm signal should be followed by a message giving essential information. A public address system used in conjunction with a fire warning system should be installed in accordance with British Standard 5839: Part 1. Reference should also be made to British Standard 7443 Specification for sound systems for emergency purposes.

Note: A sounder, or the microphone of a public address system, should not be located in such a position as to interfere with the calling of the fire brigade or the briefing of the fire brigade on arrival.

Automatic fire detection

15.12 Early detection of a fire has a significant effect on life safety as it allows persons in the premises to be alerted so that they can leave the area of the fire whilst the escape conditions are still relatively safe. The presence of an isolated area of high fire risk may not by itself necessarily require the provision of an automatic fire detection (AFD) system, unless for example, a fire could break out in an unoccupied part of the premises and not be discovered before it threatened the **means of escape** from any occupied parts of the premises.

15.13 Where a system is installed it should comply with the provisions of British Standard 5839: Part 1 for an L3 life safety system. It is also important that a commissioning and installation certificate should be provided by the installer in accordance with the British Standard. The fire warning system should be regularly maintained.

Note: The effectiveness of AFD can be considerably reduced if the credibility of the system is impaired by frequent false or unwanted calls. All such calls should be investigated to identify the reasons for the alarm and to eliminate future false calls.

15.14 If a manually operated electrical alarm system and an automatic fire alarm system are installed in the same building, they should be incorporated into a single integral system.

Notes/Amendments

Chapter 16: MEANS FOR FIGHTING FIRE

16.1 All premises should be provided with means for fighting fire for use by persons in the building, and in deciding the appropriate means, consideration should be given to the nature of the materials likely to be found there.

Classification of fires

16.2 Fires are classified in accordance with British Standard EN 2 and are as follows:-

Class A fires — Fires involving solid materials, usually of an organic nature, in which combustion normally takes place with the formation of glowing embers.

Class B fires — Fires involving liquids or liquefiable solids.

Class C fires — Fires involving gases.

Class D fires — Fires involving metals.

Class A fires

16.3 Class A fires are the most likely type of fire to occur in the majority of premises. Water, foam and multi purpose powder are the effective media for extinguishing these fires. Water and foam are usually considered to be the most suitable media and the appropriate equipment is therefore hose reels, water type extinguishers or extinguishers containing fluoroprotein foam (FP), aqueous film forming foam (AFFF) or film forming fluoroprotein foam (FFFP).

Hose reels

16.4 If hose reels are installed they should be located where they are conspicuous and always accessible eg in corridors. The hose should comply with Type 1 hose specified in British Standard 3169 and hose reel installations should conform with British Standard 5306: Part 1 and British Standard 5274.

Class B fires

16.5 Where there is a risk of fire involving flammable liquid it will usually be appropriate to provide portable fire extinguishers of foam (including

FP, AFFF and FFFP), carbon dioxide (CO_2)*, halon* or powder types. Table 1 of clause 5.3 of British Standard 5306: Part 3 gives guidance on the minimum scale of provision of various extinguishing media for dealing with a fire involving exposed surfaces of contained liquid.

*Note: * Care should be taken when using halon or CO_2 extinguishers as the fumes and products of combustion may be hazardous in confined spaces. For environmental reasons it is recommended that the provision of halon extinguishers should be avoided where other suitable extinguishing media are available. However, until suitable arrangements can be made for the disposal or banking of halons, existing fire extinguishers should remain in situ.*

Class C fires

16.6 No special extinguishers are made for dealing with fires involving gases because the only effective action against such fires is to stop the flow of gas by closing the valve or plugging the leak. There would be a risk of an explosion if a fire involving escaping gas were to be extinguished before the supply could be cut off.

Class D fires

16.7 None of the extinguishing media referred to in the preceding paragraphs will deal effectively with a fire involving such metals as aluminium, magnesium, sodium, or potassium although there is a special powder which is capable of controlling some Class D fires. Such fires should, however, only be tackled by specially trained personnel.

Portable equipment

16.8 If portable fire extinguishers are installed they should conform to British Standard 5423 and be provided and allocated to comply with clause 5.2 of British Standard 5306: Part 3.

Fire blankets

16.9 Fire blankets are suitable for some types of fire. They are classified in British Standard 6575 and are described as follows:-

(a) *light duty* – there are suitable for dealing with small fires in containers of cooking fat or oils and fires in clothing; and

(b) *heavy duty* – these are for industrial use where there is a need for the blanket to resist penetration by molten materials.

Fires involving electrical equipment

16.10 Extinguishers provided specifically for the protection of electrical risks should be of the dry powder, CO_2, or halon type. While some extinguishers containing aqueous solutions such as AFFF may meet the requirements of the electrical conductivity test of British Standard 5423 they may not sufficiently reduce the danger of conductivity along wetted surfaces such as the floor. Consequently, such extinguishers should not be provided specifically for the protection of electrical risks (see Note to paragraph 16.6).

Location of fire fighting equipment

16.11 Wherever possible, portable fire fighting equipment should be grouped to form a fire point. The fire point should be clearly and conspicuously indicated so that it can be readily identified. In premises which are uniform in layout extinguishers should, where possible, be located at the same point on each floor. If for any reason extinguishers are placed in positions hidden from direct view their position should be indicated by suitable signs. Fire safety signs are described in British Standard 5499: Parts 1 and 3. No one should have to travel more than 30 metres from the site of a fire to reach an extinguisher.

16.12 Fire extinguishers should be located in conspicuous positions, normally on escape routes ie corridors, stairways, lobbies and landings. They should normally be securely hung on wall brackets. Where this is impracticable, extinguishers should be located on a suitable base plate (not on the floor). The carrying handle of larger, heavier extinguishers should be about 1 metre from the floor but smaller extinguishers should be mounted so that the handle is positioned about 1.5 metres from the floor.

Fixed fire fighting systems

16.13 Fire fighting systems (eg sprinkler systems) will not normally be required in connection with the issue of a fire certificate. However, such systems may be necessary for the protection of life in some large and complex buildings, particularly those with a **fire safety engineering** approach. There may also be situations where other fixed suppression systems specially designed for localised protection of high fire risk areas may be justified. If a fixed system is provided for life safety purposes it should comply with the appropriate part of British Standard 5306. Any sprinkler system or other automatic fire suppression or detection installation should be linked into the fire warning system for the building.

Fire Extinguishers

The illustrations of portable extinguishers indicate the whole body in a colour which is the colour code for that particular type of extinguisher.

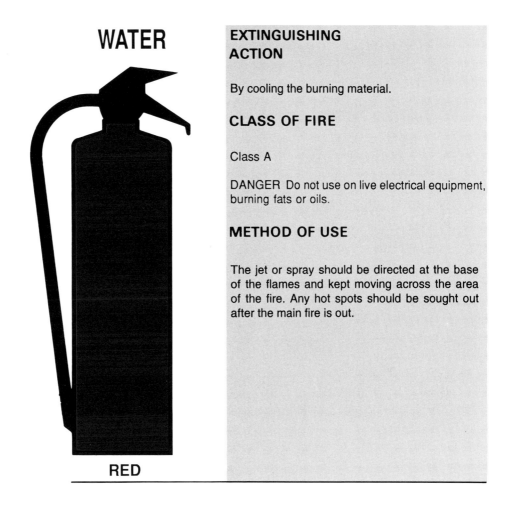

WATER

RED

EXTINGUISHING ACTION

By cooling the burning material.

CLASS OF FIRE

Class A

DANGER Do not use on live electrical equipment, burning fats or oils.

METHOD OF USE

The jet or spray should be directed at the base of the flames and kept moving across the area of the fire. Any hot spots should be sought out after the main fire is out.

FOAM
(Protein P) Type

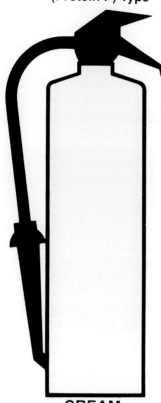

CREAM

Fluoroprotein foam (FP)

EXTINGUISHING ACTION

Forms a blanket of foam over the surface of the burning liquid and smothers the fire.

CLASS OF FIRE

Class B

DANGER Do not use on live electrical equipment.

METHOD OF USE

The jet should not be aimed directly onto the liquid. Where the liquid on fire is in a container the jet should be directed at the edge of the container or on a nearby surface above the burning liquid. The foam should be allowed to build up so that it flows across the liquid.

CREAM

Aqueous film-forming foam (AFFF) **Film-forming Fluoro-protein foam (FFFP)**

EXTINGUISHING ACTION

Forms a fire extinguishing water film on the surface of the burning liquid. Has a cooling action with a wider extinguishing application than water on solid combustible materials.

CLASS OF FIRE

Classes A & B

DANGER Some extinguishers of this type are not suitable for use on live electrical equipment.

METHOD OF USE

For Class A fires the directions for water extinguishers should be followed.

For Class B fires the directions for foam extinguishers should be followed.

POWDER

BLUE

EXTINGUISHING ACTION

Chemical inhibition of combustion.

CLASS OF FIRE

Class B

Safe on live electrical equipment although does not readily penetrate spaces inside equipment. A fire may re-ignite.

METHOD OF USE

The discharge nozzle should be directed at the base of the flames and with a rapid sweeping motion the flame should be driven towards the far edge until the flames are out. If the extinguisher has a shut-off control the air should then be allowed to clear; if the flames re-appear the procedure should be repeated.

WARNING Powder has a limited cooling effect and care should be taken to ensure the fire does not re-ignite.

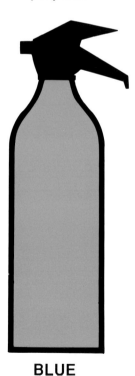

POWDER (Multi-purpose)

BLUE

EXTINGUISHING ACTION

Altering the thermal decomposition to produce non-flammable products by cooling (Class A) and chemical inhibition of combustion (Class B).

CLASS OF FIRE

Classes A & B

Safe on live electrical equipment although does not readily penetrate spaces inside equipment. A fire may re-ignite.

METHOD OF USE

The discharge nozzle should be directed at the base of the flames and with a rapid sweeping motion the flame should be driven towards the far edge until the flames are out. If the extinguisher has a shut-off control the air should then be allowed to clear; if the flames re-appear the procedure should be repeated.

WARNING Powder has a limited cooling effect and care should be taken to ensure the fire does not re-ignite.

CARBON DIOXIDE (CO_2)

BLACK

EXTINGUISHING ACTION

Displacing oxygen in the air.

CLASS OF FIRE

Class B

Safe and clean to use on live electrical equipment.

METHOD OF USE

The discharge horn should be directed at the base of the flames and the jet kept moving across the area of the fire.

WARNING CO^2 has a limited cooling effect and care should be taken to ensure that the fire does not re-ignite.

DANGER Fumes from CO^2 extinguishers can be harmful to users in confined spaces. The area should therefore be ventilated as soon as the fire has been extinguished.

HALON

GREEN

EXTINGUISHING ACTION

Vaporising liquid gas giving rapid knock down by chemically inhibiting combustion.

CLASS OF FIRE

Class B

Clean and light. Can also be used on small surface burning Class A fires. Effective and safe on live electrical equipment.

NOTE For environmental reasons it is recommended that the provision of halon extinguishers should be avoided where other suitable extinguishing media is available.

METHOD OF USE

The gas is expelled in a jet which should not be aimed into burning liquids as this risks spreading the fire. The discharge nozzle should therefore be aimed at the flames and kept moving across the area of the fire.

WARNING Halon has a limited cooling effect and care should be taken to ensure that the fire does not re-ignite.

DANGER Fumes from halon extinguishers can be harmful to users in confined spaces or if used on hot metal. The area should therefore be ventilated as soon as the fire has been extinguished.

HOSE REEL

RED

EXTINGUISHING ACTION

Water - by cooling the burning material.

CLASS OF FIRE

Class A

DANGER Do not use on live electrical equipment.

METHOD OF USE

The jet should be aimed at the base of the flames and kept moving across the area of the fire. If an isolating valve is fitted it should be opened before the hose is unreeled.

RED

Light duty

Heavy duty

EXTINGUISHING ACTION

Smothering

CLASS OF FIRE

Classes A & B

Suitable for burning clothing and small fires involving burning liquids.

Suitable for industrial use.
Resistant to penetration by molten materials.

METHOD OF USE

The blanket should be placed carefully over the fire and the hands shielded from the fire. Care should be taken that the flames are not wafted towards the user or bystanders.

British Standard 5423 recommends that extinguishers should be
(a) predominantly red with a colour coded area; (b) predominantly colour coded; or (c) of self-coloured metal with a colour coded area.

Notes/Amendments

Chapter 17: INSTRUCTION AND TRAINING IN FIRE PRECAUTIONS

17.1 Wherever premises require a fire certificate, the fire authority should ensure that the occupier is aware of the need to train staff, and all who may work on the premises, in the action which should be taken in the event of fire.

Note: The fire authority must inform an applicant for a fire certificate of the occupier's (owner's) duty to train staff in the action to be taken in the event of a fire, pending disposal of the application (see paragraph 3.3).

17.2 Training should be specific to the particular premises and all staff should receive training at sufficiently regular intervals to ensure that existing members of staff are reminded of the action to take, and that new staff are made aware of the fire routine for the premises. Training should be given at least once in each period of 12 months but in some circumstances where there is a high turnover of staff, or where there is a high fire risk, or material changes which significantly affect the fire safety arrangements, training may need to be more frequent. Instruction and training should be based on written procedures and should be appropriate to the duties and responsibilities of the staff.

17.3 It is particularly important that all staff (including those casually employed) should be shown the **means of escape** and told about the fire routine as soon as possible after they start work. There is also a need to ensure that occasional workers, those on shift duties and others who work in the premises are similarly instructed. Special consideration should be given to any employees with language difficulties or with any disabilities which may impede their understanding of the information.

17.4 Instruction should be given by a competent person and the following topics, where appropriate, should be covered in each training session with practical exercises where possible:-

(a) the action to take on discovering a fire;

(b) how to raise the alarm, and the procedures this sets in motion;

(c) the action to be taken upon hearing the fire alarm;

(d) the procedures for alerting members of the public including, where appropriate, directing them to exits;

(e) the arrangements for calling the fire brigade;

(f) the evacuation procedure for the premises to an assembly point at a **place of safety**;

(g) the location and use of fire fighting equipment;

(h) the location of the escape routes, including those not in regular use;

(i) how to open all escape doors;

(j) the importance of keeping **fire doors** closed;

(k) how to stop machines and processes and isolate power supplies where appropriate;

(l) the reason for not using lifts (other than those specifically provided or adapted for use by people with disabilities in accordance with BS 5588: Part 8); and

(m) the importance of general fire precautions and good housekeeping (see Chapter 21).

17.5 There will also be staff involved in certain activities who will need instruction in their specific duties in the event of a fire. These could include:-

kitchen staff
engineering and maintenance staff
receptionists and
telephonists.

17.6 In addition to the subjects covered in paragraph 17.4, supervisory staff and those with a particular responsibility in case of fire should be given additional instruction and training. These will include:-

heads of departments
supervisors
security staff
fire marshals.

17.7 A fire drill should be carried out at least once and preferably twice a year simulating conditions in which one or more of the escape routes from the building is obstructed. During these drills the fire alarm should be operated by a member of staff who is told of the supposed outbreak, and thereafter the fire routine should be rehearsed as fully as circumstances allow.

Note: Normally, advance warning should be given of the fire drill as the main purpose of having the drill is to ensure that all persons participating are familiar with the correct procedure to be followed. Every opportunity should be taken to learn lessons from any fire drill and to reinforce staff training where appropriate.

17.8 The training and instruction given should be recorded in a log book or other suitable record and should be available for inspection. The following are examples of matters which should be included in such a record:-

(a) the date of the instruction or exercise;

(b) the duration;

(c) the name of the person giving the instruction;

(d) the names of the persons receiving the instruction; and

(e) the nature of the instruction, training or drill.

17.9 In all premises one person should be responsible for organising fire instruction and training and in larger premises a person or persons should be nominated to co-ordinate the actions of the occupants in the event of a fire.

Fire action notices

17.10 Printed notices should be displayed at conspicuous positions in the building stating in concise terms the action to be taken upon discovering a fire or on hearing the fire alarm. The notices should be permanently fixed in position and suitably protected to prevent loss or defacement. The content of the notice will normally be specified in the fire certificate. Written instructions may be supplemented by advice in pictogram form.

Notes/Amendments

PART III
ADVICE WHICH A FIRE AUTHORITY MAY BE ASKED TO PROVIDE TO OCCUPIERS OR OWNERS OF FACTORY, OFFICE, SHOP OR RAILWAY PREMISES

Chapter 18: THE RESPONSIBILITIES OF MANAGEMENT

Responsibilities imposed by the Fire Certificate

18.1 In certificated premises it is the occupier/owner who has responsibility for ensuring, among other things, the maintenance of:-

(a) adequate **means of escape** (see Chapter 14);

(b) a system whereby staff can be alerted to the presence of a fire (see Chapter 15); and

(c) suitable means for fighting fire (see Chapter 16).

This responsibility extends to ensuring that staff are trained in the action to take in the event of fire and know what they should do if they discover a fire, or if the fire alarm is sounded (see Chapter 17).

18.2 The fire certificate will normally require details of staff training, fire drills and the testing and maintenance of the fire warning system, **emergency escape lighting** and fire fighting equipment to be recorded. This record should be kept on the premises and be available for inspection.

Fire safety management

18.3 The occupier needs to encourage and instil a positive awareness of fire safety in order to minimise the risk of fire breaking out in the premises and protect the lives of his or her staff and members of the public. Comprehensive guidance is available to managers in a document entitled "Fire Safety at Work", published by HMSO (reference ISBN 0 11 340905 2).

18.4 The occupier's responsibility is normally a personal one but it may be delegated if a person has been employed for that purpose. Whatever the size of the premises there should be no doubt where this responsibility lies and if the occupier is absent, some other person should have the authority to act in his or her place.

18.5 The occupier, or individual appointed as having responsibility for fire safety should be able to:-

(a) assess the risk of fire in the premises (see paragraphs 13.5 to 13.8);

(b) take action to minimise the likelihood of a fire occurring (see Chapter 21);

(c) take action to limit the spread of fire (see Chapter 20);

(d) ensure that staff/visitors are adequately informed on what to do in case of fire; and

(e) establish procedures to be followed in the event of fire.

18.6 It is prudent for the occupier to have in readiness a plan of action to be taken, following any fire which may occur, to include details of salvage companies, architects, builders, contractors etc. The contingency arrangements should be kept up to date.

Notes/Amendments

Chapter 19: ASSISTING PEOPLE WITH DISABILITIES

General

19.1 In certificated premises the occupier or owner may seek advice about fire safety provisions for people with disabilities. Non-technical advice is available in 'Fire Safety at Work" published by HMSO (reference ISBN 0 11 340905 2).

19.2 British Standard 5588: Part 8 Code of practice for means of escape for disabled people, provides guidance on the measures to be incorporated into new buildings or existing buildings which have been altered, to enable the safe evacuation of people with disabilities in the event of fire. It is accepted that it may not always be possible to fully comply with the code in existing buildings and in these circumstances alternative ways of meeting its objectives should then be sought.

19.3 In most cases, the number of people with disabilities will be small in relation to the total number of employees and no specific additions to the **means of escape** will be necessary. However, the fire routine procedure should ensure that the needs of any staff who are disabled are adequately catered for and nominated members of staff should be given the specific responsibility of assisting such staff in the event of fire.

19.4 Management should draw up the fire routine for the premises in consultation with any staff who have disabilities and by obtaining any necessary specialist advice.

Assisting the less able-bodied

19.5 If people use wheelchairs, or can move about only with the aid of a stick or crutches, their disability is obvious. However disabilities are less obvious for people who have had strokes or heart attacks, those who are arthritic or epileptic, and those with poor sight or hearing. There are also many people, such as those with broken limbs and other injuries, and women in the late stages of pregnancy whose condition affects their mobility. Elderly people and young children may also require special consideration.

Assisting wheelchair users and people with impaired mobility

19.6 In drawing up an evacuation plan for the premises management should consider how wheelchair users and people with impaired mobility can be assisted. Lifts should not be used in the event of a fire unless they are specially designed or adapted for the evacuation of the disabled as described in British Standard 5588: Part 8. Where stairs need to be negotiated and there is a likelihood that members of staff may have to be carried, managers should consider training able-bodied members of staff in the correct methods of doing so. Advice on lifting and carrying people with disabilities may be obtained from the fire brigade, the ambulance service, the British Red Cross Society, the St John Ambulance Brigade, the St Andrew's Ambulance Association or certain of the access or disability organisations listed at paragraph 19.14.

Assisting people with impaired vision

19.7 People with impaired vision or colour perception may experience difficulty in recognising fire safety signs. However, many people are able to read print if it is sufficiently large and well designed with a good clear typeface. Advice about the design and siting of signs and notices may be obtained from the Royal National Institute for the Blind or the National Federation of the Blind of the UK. Good lighting and use of simple colour contrasts can also help visually impaired people find their way around.

19.8 Managers should ensure that any employees with impaired vision familiarise themselves with escape routes, especially those which are not in general daily use. Where possible, management should arrange for a normally sighted person to work near any employee with impaired vision to warn and reassure them in the event of a fire. A member of staff should also accompany the individual along the escape route and ensure that he or she is not abandoned after leaving the building but accompanied to a **place of safety**. It is recommended that the sighted person should lead, inviting the other person to grasp their elbow or shoulder lightly as this will enable the person being assisted to walk half a step behind and thereby gain information about doors and steps, etc. Similar assistance should be offered to guide dog owners, but control of the dog should remain with the owner.

Assisting people with impaired hearing

19.9 Although people with impaired hearing may experience difficulty in hearing a fire alarm, they may not be completely insensitive to sound. Many people with severe impairment have sufficiently clear perception of some types of conventional audible alarm signals to require no special provision. Managers should however ensure that where a member of staff does have difficulty, a colleague is given the responsibility of alerting the individual concerned.

19.10 In certain situations, such as premises where profoundly deaf people form the majority of employees, alternative types of alarm signal may be necessary eg lights or other visual signals, vibrating devices, or sound signals within carefully selected frequency bands. Management may obtain technical advice on the selection of suitable devices from the Royal National Institute for Deaf People. As alternative alarm signals may have unwanted side effects they should only be installed following adequate consultation between management and employees.

19.11 Induction loop systems, used in some premises for audio communication with people with suitable hearing aids, are not considered acceptable means of alerting hearing-impaired people in the event of fire. Where such systems are in normal use, however, they may be used to supplement the alarm.

Assisting people with learning difficulties or mental illness

19.12 Employers should ensure that any staff with learning difficulties or mental illness are told what they should do in the event of fire. Arrangements should be made to ensure that they are assisted and reassured in the event of fire and accompanied to a **place of safety**. They should not be left unattended. Advice may also be sought from MENCAP, or from local residential or day services for people with learning difficulties.

Use of lifts

19.13 Disabled people may rely on a lift as a **means of escape** only if it is an evacuation lift or a firefighting lift operated under the direction and control of management using an agreed evacuation procedure. The recommendations in paragraph A.3 of Appendix A to British Standard 5588: Part 8 should be followed.

Sources of advice

19.14 Names and addresses of organisations representing people with disabilities can be found in Yellow Pages. For convenience, some of the principal organisations are listed below.

1. Access Committee for England, 25 Mortimer Street, London, W1N 8AB

2. Association of Access Officers, c/o Wrekin District Council, PO Box 212, Malinsee House
   Telford, Shropshire TF3 4LL

3. British Council of Organisations of Disabled People, de Bradelei House, Chapel Street, Belper, Derbyshire DE56 1AR

4. Centre for Accessible Environments, 35 Great Smith Street, London, SW1P 3BJ

5. Disability Scotland, Princes House, 5 Shandwick Place, Edinburgh EH2 4RG.

6. Disabled Living Foundation, 380–384 Harrow Road, London W9 2HU.

7. Joint Committee on Mobility for Disabled People, Woodcliff House, 51A Cliff Road, Weston-Super-Mare, Avon BS22 9SE

8. MENCAP, National Centre, 123 Golden Lane, London EC1Y 0RT

9. MIND, 22 Harley Street, London W1N 2ED

10. National Federation of the Blind of the United Kingdom, Unity House, Smyth Street, Westgate, Wakefield, West Yorkshire WF1 1ER.

11. RADAR (Royal Association for Disability and Rehabilitation), 25 Mortimer Street, London W1N 8AB

12. Royal National Institute for the Blind, 224 Great Portland Street, London W1N 6AA/10 Magdala Crescent, Edinburgh EH12 5BF

13. Royal National Institute for Deaf People, Science and Technology Unit, 105 Gower Street, London WC1E 6AH/ 9 Clairmont Gardens, Glasgow G3 7LW.

14. Wales Council for the Disabled, Llys Ifor, Crescent Road, Caerphilly, Mid Glamorgan, CF8 1XL.

Notes/Amendments

Chapter 20: FLOOR COVERINGS, FURNITURE, FURNISHINGS AND SYNTHETIC MATERIALS, ARTIFICIAL AND DRIED FOLIAGE

20.1 Whilst there are no nationally specified standards in relation to the flammability of furniture etc. in the premises covered by this guide fire authorities may wish to have regard to the following matters in giving advice (see also Appendix B).

Floor Coverings

20.2 Some floor coverings, when involved in fire, may react to produce large volumes of heat and smoke although the surface spread of flame may be relatively slow. The possibility that the floor coverings may present a hazard to the **means of escape** should not be overlooked and should feature in the overall assessment of suitability of surfaces, along with those for walls and ceilings to **protected routes**. If new floor coverings are to be installed, they should comply with British Standard 5287 as conforming to the low radius of fire spread (up to 35 mm) when tested in accordance with British Standard 4790.

Furniture, furnishings and synthetic materials

20.3 Furniture, furnishings and synthetic materials which are easily ignited, demonstrate rapid spread of flame characteristics or produce large quantities of smoke or toxic fumes should be a factor in determining the acceptability of escape routes and in particular of **protected routes.**

Artificial and dried foliage

20.4 It is not possible to assess dried or artificial foliage in terms of flame retarded fabrics using formal laboratory test methods. It is, however, recommended that these and similar items should have been subjected to ignition tests using small flaming sources (see also Appendix B).

Notes/Amendments

Chapter 21: GOOD HOUSEKEEPING AND THE PREVENTION OF FIRE

General

21.1 The fire authority may be asked to give advice on good housekeeping and fire prevention. The paragraphs which follow are based on Chapter 7 of "Fire Safety at Work", published by HMSO (reference ISBN 0 11 340905 2).

21.2 The need for good housekeeping and sensible fire precautions cannot be over-emphasised as these practices will reduce the possibility of a fire occurring. Poor housekeeping, carelessness and neglect not only make the outbreak of fire more likely but will inevitably allow a fire to spread more rapidly.

The causes of fire and common fire hazards

21.3 Most fires can be prevented. The attention of staff should be drawn to the common fire hazards and causes of fire which include:-

(a) faulty electrical wiring; plugs and sockets which are in poor condition, overloaded or inadequately protected by fuses or other devices;

(b) electrical equipment left switched on when not in use (unless it is designed to be permanently connected);

(c) smoking in unauthorised or high fire risk areas and the careless disposal of smoking materials;

(d) accumulation of rubbish, paper or other materials which can easily catch fire;

(e) careless use of portable heaters;

(f) excess or careless storage of flammable substances and combustible materials;

(g) obstructing the ventilation of heaters, machinery and electrical appliances including office equipment;

(h) inadequate cleaning of work areas and poorly maintained equipment;

(i) inadequate supervision of cooking and other work activities; and

(j) carelessness by contractors or maintenance workers.

List of fire precautions

21.4 The individual responsible for fire safety (the occupier, owner or manager) should inspect the premises to identify any potential fire hazards. It is then suggested that a list of fire precautions should be drawn up under suitable headings which may include the following:-

(a) electrical equipment and installations;

(b) smoking and the provision of ashtrays;

(c) kitchens;

(d) building and maintenance work;

(e) combustible rubbish and waste;

(f) furniture and furnishings, (see also Chapter 20 and Appendix B); and

(g) unoccupied areas.

Electrical equipment

21.5 All staff should be instructed in the correct use of electrical equipment, in the recognition of faults and in how to report faults to supervisors. The use of multiple adaptors in electrical socket outlets and coiled extension leads should be avoided as this may lead to overheating. The occupier/owner should make sure that all electrical repairs are carried out by a qualified electrician.

Smoking and the provision of ashtrays

21.6 The occupier should have a positive policy regarding smoking which identifies the areas where smoking is permitted. As the careless disposal of smoking materials is one of the main causes of fire the occupier should ensure that, in areas where they allow smoking, staff have a plentiful supply of ashtrays and these are emptied regularly. Ashtrays should not be emptied into containers which can be easily ignited, nor should their contents be disposed of with general rubbish. The occupier should not allow smoking in store rooms, kitchens, and other utility areas.

Kitchens

21.7 The misuse of grill trays, frying pans, deep fat frying equipment and microwave ovens can cause fires. The occupier should therefore ensure that staff are instructed in the correct use of equipment and know how to prevent a fire occurring (see Chapter 17).

21.8 Ideally main electrical switches and gas stopcocks in a kitchen should be positioned on an exit route and be clearly indicated. Switches designed to isolate the extraction fans, in order to prevent flames spreading through extraction systems, should be similarly located. Extraction fans should normally be linked into a fire detection system so that the fans are automatically closed down in the event of fire.

Building and maintenance work

21.9 Many serious fires occur during building and maintenance work and any such work should therefore be closely supervised (see also Chapter 13). In larger and complex premises a hot work permit system should be used. The occupier should ensure that any location where hot work (such as welding or using a blow lamp or torch) is to take place is examined to make sure that all material which could be easily ignited has either been removed or has been suitably protected against heat and sparks. Suitable extinguishers should be readily available. Any areas where hot work is undertaken should be frequently inspected during the first 30 minutes after the work is completed, and again 30 minutes later to ensure that materials are not smouldering.

21.10 Hazardous substances such as flammable adhesives, cleaning materials and paints (others are described in Chapter 13) should be securely stored in a well ventilated area and when not in use kept separate from other materials. When using flammable adhesives and cleaning fluids rooms should be well ventilated and free from sources of ignition. Gas cylinders should be stored securely outside the building.

21.11 At the end of the working day a check should be made to ensure that all flammable substances, combustible material and equipment are safe and that no fires can start accidentally. Special care should be taken when restoring gas and electricity supplies to ensure that equipment has not inadvertently been left switched on.

Combustible rubbish and waste

21.12 Rubbish should not be stored, even as a temporary measure, in escape corridors, stairways or lobbies. Accumulations of waste should be avoided and all rubbish and waste should be removed at least daily and suitably stored clear of the building.

Furniture and furnishings

21.13 As the fibre and cellular foam fillings in most upholstered furniture can be ignited by smoking materials the occupier should ensure that staff check regularly to make sure that there are no tears or rips which have resulted in the filling material being exposed.

Unoccupied areas

21.14 Parts of the premises which are not normally occupied, such as basements, store rooms and any area where a fire could grow unnoticed should be regularly inspected and kept clear of non-essential flammable substances and combustible materials (see Chapter 13). Care should also be taken to protect such areas against entry by unauthorised persons.

Note: Factories may contain significant fire hazards associated with both the storage of materials and the processes carried out. Advice regarding these matters is available from HSE or local HSE Inspectors.

Check list

21.15 Although fire precautions are mainly commonsense, staff need to know what to look for. The occupier should therefore draw up a checklist to ensure that:-

(a) the **means of escape** are well signposted and kept clear of obstruction at all times;

(b) internal **fire doors** are clearly labelled and any self-closing devices are kept in working order;

(c) all fire exit doors can be easily and immediately opened from the inside without the use of a key;

(d) there are no obstructions, apparent defects or damage to fire alarm call points, fire detectors or alarm sounders;

(e) the fire-fighting equipment is in good order, unobstructed and in place;

(f) all electrical equipment is fitted with fuses of the correct size and type, and that lengths of flexible cable are kept to the minimum; that cables are run only where damage is unlikely and never under floor coverings or through doorways;

(g) material which could readily catch fire is not left near to a source of heat;

(h) flammable substances or combustible materials are suitably located and kept in appropriate quantities;

(i) there are adequate facilities for the disposal of smoking materials;

(j) all furnishings are in good condition;

(k) there is no accumulation of rubbish, waste paper or other materials which could catch fire. (Such a check is particularly important when part of the premises has been used for a seminar or exhibition);

(l) decorative materials used at festive or social gatherings are not readily ignitable; that decorations are not attached to lights or heaters, and that they do not obscure fire safety notices and **emergency escape lighting**;

(m) heating appliances are fixed in position at a safe distance from any combustible materials and are adequately guarded; and

(n) open fires are protected with fixed guards to prevent the risk of sparks igniting materials in the close vicinity.

Note: (a), (b), (c), (d) and (e) will normally be included on the fire certificate.

21.16 Although the guidance in this chapter is about protecting persons from the risk of fire many of the precautions recommended will also protect the building and its contents whilst the premises are unoccupied. It is therefore suggested that *at the end of the day's activities* a full check is carried out to make sure that:-

(a) the building is secured against unauthorised entry;

(b) all doors are closed, including those held open during the day by automatic door release units;

(c) electrical equipment not in use is switched off and, where appropriate, unplugged;

(d) smoking materials are not left smouldering;

(e) all rubbish and waste is removed; and

(f) all combustible materials and flammable substances are safely stored.

21.17 Effective arrangements should be made to ensure that any deficiency found during the checks carried out in accordance with paragraph 21.15 is speedily rectified.

21.18 Staff should be encouraged to bring any potential fire risk to the attention of supervisors.

Notes/Amendments

Appendix A

REFERENCES TO BRITISH, EUROPEAN AND INTERNATIONAL STANDARDS

1. Where references are made to British and European Standards in this guide, these are technical standards published by the British Standards Institution which are in force at the time of publication. Where premises are brought into use after the publication of this guide, the relevant British Standards will be those current at the time the work is undertaken.

2. It should be noted that where a current British Standard is superseded by an European Standard published by the European Committee for Standardisation (CEN), the existing British Standard will either be withdrawn or the CEN Standard will be published by the British Standards Institution in their BS EN series. In the case of any such European Standard being cited in the Essential Requirements of a Directive, compliance with that Standard becomes mandatory for the purposes of that Directive.

3. International Standards published by the International Organisation for Standardisation (ISO) or the International Electrotechnical Committee (IEC) are normally published in the BS General Series, with reference being made to the ISO or IEC number.

British Standards mentioned in the guide are listed below:-

BS 476:-	FIRE TESTS ON BUILDING MATERIALS AND STRUCTURES:
Part 6	Method of test for fire propagation of products.
Part 7	Method for classification of the surface spread of flame of products.
Part 21	Methods for determination of the fire resistance of loadbearing elements of construction.
Part 22	Methods for determination of the fire resistance of non-loadbearing elements of construction.
Part 23	Methods for determination of the contribution of components to the fire resistance of a structure.
Part 24	Method for determination of the fire resistance of ventilation ducts
BS 3169:-	SPECIFICATION FOR FIRST AID REEL HOSES FOR FIRE FIGHTING PURPOSES
BS 4790:-	METHOD FOR DETERMINATION OF THE EFFECTS OF A SMALL SOURCE OF IGNITION

	ON TEXTILE FLOOR COVERINGS (HOT METAL NUT METHOD)
BS 5266:-	EMERGENCY LIGHTING:
Part 1	Code of practice for the emergency lighting of premises other than cinemas and certain other specified premises used for entertainment.
BS 5274:-	SPECIFICATION FOR FIRE HOSE REELS (WATER) FOR FIXED INSTALLATIONS
BS 5287:-	SPECIFICATION FOR ASSESSMENT AND LABELLING OF TEXTILE FLOOR COVERINGS TESTED TO BS 4790
BS 5306:-	FIRE EXTINGUISHING INSTALLATIONS AND EQUIPMENT ON PREMISES:
Part 1	Hydrant systems, hose reels and foam inlets.
Part 2	Specification for sprinkler systems.
Part 3	Code of practice for selection, installation and maintenance of portable fire extinguishers.
Part 4	Specification for carbon-dioxide systems
Part 5	Section 5.1 Specification for Halon 1301 total flooding system.
	Section 5.2 Halon 1211 total flooding system
Part 6	Section 6.1 Specification for low expansion foam systems.
	Section 6.2 Specification for medium and high expansion foam systems.
Part 7	Specification for powder systems.
BS 5395:-	STAIRS, LADDERS AND WALKWAYS
Part 2	Code of practice for the design of helical and spiral stairs.
BS 5423:-	SPECIFICATION FOR PORTABLE FIRE EXTINGUISHERS
BS 5438:-	METHODS OF TEST FOR FLAMMABILITY OF VERTICALLY ORIENTED TEXTILE FABRICS AND FABRIC ASSEMBLIES SUBJECTED TO A SMALL IGNITING FLAME

BS 5499:-	**FIRE SAFETY SIGNS, NOTICES AND GRAPHIC SYMBOLS:-**
Part 1	Specification for fire safety signs.
Part 3	Specification for internally-illuminated fire safety signs.
BS 5588:-	**FIRE PRECAUTIONS IN THE DESIGN AND CONSTRUCTION OF BUILDINGS:**
Part 2	Code of practice for shops.
Part 3	Code of practice for office buildings.
Part 4	Code of practice for smoke control in protected escape routes using pressurization.
Part 6	Code of practice for places of assembly.
Part 7	Code of practice for atria (under preparation).
Part 8	Code of practice for means of escape for disabled people.
Part 9	Code of practice for ventilation and air conditioning ductwork.
Part 10	Code of practice for shopping complexes.
Part 11	Code of practice for workplaces (under preparation).
BS 5651:-	**METHOD FOR CLEANSING AND WETTING PROCEDURES FOR USE IN THE ASSESSMENT OF THE EFFECT OF CLEANSING AND WETTING ON THE FLAMMABILITY OF TEXTILE FABRICS AND FABRIC ASSEMBLIES**
BS 5725:-	**EMERGENCY EXIT DEVICES**
Part 1	Specification for panic bolts and panic latches mechanically operated by a horizontal push-bar.
BS 5839:-	**FIRE DETECTION AND ALARM SYSTEMS FOR BUILDINGS:**
Part 1	Code of practice for system design, installation and servicing.
Part 2	Specification for manual call points.
Part 3	Specification for automatic release mechanisms for certain fire protection equipment.
BS 5852:-	**METHODS OF TEST FOR ASSESSMENT OF THE IGNITABILITY OF UPHOLSTERED SEATING**

	BY SMOULDERING AND FLAMING IGNITION SOURCES
BS 5867:-	SPECIFICATION FOR FABRICS FOR CURTAINS AND DRAPES
Part 2	Flammability requirements.
BS 6262:-	CODE OF PRACTICE FOR GLAZING FOR BUILDINGS
BS 6575:-	SPECIFICATION FOR FIRE BLANKETS
BS 6807:-	METHODS OF TEST FOR THE IGNITABILITY OF MATTRESSES, DIVANS AND BED BASES WITH PRIMARY AND SECONDARY SOURCES OF IGNITION
BS 7176:-	SPECIFICATION FOR RESISTANCE TO IGNITION OF UPHOLSTERED FURNITURE FOR NON- DOMESTIC SEATING BY TESTING COMPOSITES
BS 7443	SPECIFICATION FOR SOUND SYSTEMS FOR EMERGENCY PURPOSES
BS 8214:-	CODE OF PRACTICE FOR FIRE DOOR ASSEMBLIES WITH NON-METALLIC LEAVES

European Standards mentioned in the guide are listed below:-

BS EN 2	CLASSIFICATION OF FIRES

Copies of British Standards may be obtained from the Sales Department, British Standards Institution, Linford Wood, Milton Keynes, MK14 6LE (telephone 0908 221166).

Notes/Amendments

Appendix B

TECHNICAL STANDARDS FOR UPHOLSTERED FURNITURE, ARTIFICIAL FOLIAGE, TREES, SHRUBS AND FLOWERS

Upholstered furniture

1. It is recommended that upholstered furniture should only contain those filling materials specified in the Furniture and Furnishings (Fire) (Safety) Regulations 1988; and that invisible parts of permanent covers should comply with the provisions of the Furniture and Furnishings (Fire) (Safety) Amendment Regulations 1989. The furniture should satisfy a minimum standard Ignition Source 0 (cigarette) and Ignition Source 5 (timber crib) of British Standard 5852. In addition, where the cover material, or any barrier fabric, has been treated chemically to impart a degree of flame retardance, it should be subjected to the water soak test, as advocated in the Furniture and Furnishings (Fire) (Safety) Regulations 1988. The new standards should be applied when refurbishment and replacement takes place.

2. Certification of upholstered furniture to British Standard 5852, in every instance, would be unnecessarily burdensome and not in keeping with the precedent set by the Department of Trade and Industry Regulations which allows predictive testing in the case of domestic upholstered furniture.

3. British Standard 7176: Specification for Resistance to ignition of upholstered furniture for non-domestic seating by testing composites, specifies the direct testing of the actual composite in use but also allows predictive testing for small orders (less than 200 identical units in any 12 month period), where a "worst case" composite is tested. This route to compliance is described in Appendix A of the British Standard which suggests a scheme whereby the manufacturer can test the cover over a standard substrate and test the filling under a standard cover. The label attached to the furniture would then indicate which route to compliance had been followed ie direct or predictive testing.

4. Fire authorities should advise owner/occupiers that either route to compliance is acceptable but that direct testing is the definitive performance and that in some situations direct testing would be required regardless of the number of items to be provided.

Artificial foliage, trees, shrubs and flowers

5. It is recommended that these and similar items should have been subjected to ignition tests using small flaming sources, as follows:-

> a suitable small flaming ignition source, the match equivalent butane flame Ignition Source 1 specified in British Standard 5852, should be applied to the treated leaves, flowers, etc. of the sample for 20 seconds. Ignition is acceptable during the application of the igniting flame, but on its removal, flaming, whilst continuing locally, should

not spread beyond the area first ignited. Care should be taken to test each component of artificial shrubs or trees, not only the leaves and flowers.

6. As it has been found difficult to totally inhibit the production of flaming molten droplets or debris from the solid plastics parts of artificial foliage such as stems, fire authorities should consider such factors as:-

(a) location;

(b) ease of access by members of the public; and

(c) the overall amounts of artificial foliage present.

7. All artificial and dried foliage used for decorative purposes in public areas should be flame retardant treated. As flame retardant treatments can be adversely affected by contact with moisture (as this can cause recrystallisation on surfaces), periodic retreatment may be required. Fire Authorities should advise occupiers/owners to consider a policy of reappraisal for such treated items.

8. Dried flowers and grasses should not be sprayed with hair lacquer or other like substances, as such treatment will only enhance the ease of ignition and rate of fire spread.

Notes/Amendments

INDEX

	Paragraph(s)
Access	
— and inner rooms	14.18–14.19, diagram 5
— and means of escape for people with disabilities	Chapter 19
Accommodation stairways	14.56, 14.58
AFD (Automatic Fire Detection)	15.12–15.14
Alarms	
— action on hearing	17.4, 17.10
— action on sounding	17.4, 17.7
— audible	15.11
— call points	15.8
— false	Note to 15.13
— installation	15.2, 15.11, 15.13
— manually operated	15.3
— people with disabilities	17.3
— phased	15.6
— staff	15.7
— staged	15.4–15.6
Alteration, material, meaning of	3.26
Alternative routes from a room or storey	14.8
Angle of divergence	14.8, diagrams 1 and 2
Appeal, rights of	Chapter 7
Application for a fire certificate	3.1–3.3
— deemed to have been withdrawn	3.5
Artificial and dried foliage	20.4, Appendix B
Assessment	
— high fire risk	13.8–13.11, 13.13
— low fire risk	13.5
— means of escape	14.11
— normal fire risk	13.6
Assisting people with disabilities	Chapter 19
Audible alarms by intercommunication or public address equipment	15.11
Automatic	
— door releases	14.85
— fire detection	13.3, 13.13(h), 15.12–15.14
— fire suppression	13.3, 13.13(h)
— lowering lines	14.69
Balconies	14.78
Basements	
— fire resistance of floor immediately over	Table A & Notes, 14.51–14.55
— stairways from	14.51–14.55, diagrams 18 & 19
Basic principles for means of escape	14.3–14.8
Blind people	19.7–19.8
Bridges	14.78
British, European and International Standards	Introduction, Appendix A
Building	
— single stairway	14.36–14.39, diagrams 7 to 12
— with two or more stairways	14.40–14.50, diagrams 13–17
Building Regulations	Introduction, Chapter 10
Building Standards (Scotland) Regulations	Introduction, 10.2
By-pass route avoiding a stairway enclosure	14.43, 14.78, diagram 13

Call points	15.8
Cavities	
— fire stopping of	12.7
— not readily apparent	12.8
Cellular foam filled furniture	Note to 13.13, 21.13
Classes for surface spread of flame	12.3–12.5, Table B
Classification of fires	16.2, 16.4, 16.6–16.8
Code of practice for means of escape for people with disabilities	19.2
Common fire hazards and causes of fire	21.3
Consideration of exemption	3.4
Consultation	Chapter 11
— with fire authority about fire alarms	15.5–15.6
— with staff who have disabilities	19.4
Corridors	14.27–14.32
Criminal Justice Act 1982	6.2–6.3
Criminal Procedure (Scotland) Act 1975	6.3–6.4
Crown premises	Chapter 9
Cupboards: doors to	14.30, 14.97–14.98
Dead-end, initial	14.17 and diagram 4
Deaf/hearing impaired people	19.9–19.11
Decorations	
— artificial/dried foliage	21.15(1), Appendix B
Defence	
— ignorance may (not) be acceptable as	6.7
— of due diligence	6.11
Directional signs	14.93–14.95
Disabled people	Chapter 19
— code of practice of means of escape for	19.2
— door closing devices	14.85
— evacuation procedure for using lifts	19.13
— sources of advice concerning	19.14
Disabled staff	
— consultation with	19.4
Disclosure of information	3.10
Display/exhibition areas	13.13
Display materials and display boards	12.6
Distance of travel	14.13–14.55, Table D and diagrams 3 & 4
Doors	
— automatic releases for	14.85
— existing, modification of	Table A & Notes, 14.31
— fastenings on	14.86–14.89
— fire	Table A & Notes, 14.31, 14.52–14.53, 14.83–14.84, 14.97–14.98, 17.4, diagrams 18 and 19, 21.15
— fire resistance of	Table A & Note 7, 13.13
— goods delivery	14.81
— hold open devices	14.85
— intercommunicating	14.43, 14.77
— loading	14.81
— on escape routes	14.83
— opening onto stairway or corridor	14.48, 14.50
— revolving	14.90
— self-closing devices for	14.84

Doors continued
- signs on — 14.96–14.99
- sliding — 14.92
- smoke seals — 14.31, 14.83
- sub-dividing a corridor — 14.31, diagram 6
- to balconies, bridges and walkways — 14.78
- to cupboards, service ducts, vertical shafts — 14.30, 14.84
- toilet — Table A & Notes
- up and over — 14.81
- vision panel in — Table A & Notes, 14.18–14.19, diagram 5, 14.83
- wicket — 14.80

Dried grass, flowers and foliage — 20.4 & Appendix B
Ducts, ventilation — Table A, 12.7–12.8, 14.84
Due diligence — 6.11

Emergency escape lighting — 14.101–14.102
Emergency plan — 18.6
Enclosure of stairways — 14.42–14.50 and diagrams 14, 15 and 16
Escalators/travolators — 14.57–14.59
Escape
- devices — 14.69
- from tall items of plant — 14.70
- hatches — 14.76–14.77
- ladder, portable — 14.67
- lighting — 14.78, 14.100–14.102

Evacuation
- general arrangements for/plan — Chapters 17 & 19
- lifts — 19.13
- procedure for people with disabilities — Chapter 19

Exemption from the requirement to have a fire certificate — Chapter 4
Exit and directional signs — 14.93–14.95
Exits
- number/width/siting of — 14.5, 14.20–14.24
- roof — 14.79
- used by fewer than 5 persons — 14.22, 14.80
- window — 14.82, diagram 22

Explosives Act 1875 — 8.2
Extension, material, meaning of — 3.26
Extinguishers — Chapter 16

Fastenings on doors and windows — 14.82(f) 14.86–14.89
Fines — Chapter 6
Fire action notices — 17.10
Fire alarms — Chapter 15, 17.4, 19.9
Fire blankets — 16.9
Fire certification procedure — Chapters 3, 7 and 9
Fire detection, automatic — 15.12–15.14
Fire drills — Chapter 17
Fire exit signs — 14.93–14.95
Fire extinguishers — Chapter 16
Fire fighting equipment — Chapter 16
Fire inspector — 3.7–3.10, 6.7, Chapter 9
Fire instruction and drills — Chapter 17

Fire precautions	21.4–21.18
Fire Precautions Act 1971	Part I
Fire Precautions legislation	Chapter 1
Fire Precautions (Places of Work) Regulations	Introduction
Fire resistance and surface furnishes of walls, ceilings and escape routes	Chapter 12
Fire risk and associated life risk	Chapter 13
Fire Safety and Safety of Places of Sport Act 1987	Introduction, 1.4–1.5
Fire safety engineering	14.9–14.10
Fire safety management	Chapter 18
Fire safety signs and graphic symbols	14.93–14.99
Fire Service Inspectorate as fire authority	1.1 and Chapter 9
Fire/smoke spread	12.7–12.8
Fire/smoke stopping	12.7
Fire training	Chapter 17
Fire warning signals	15.9–15.10
Fires involving burning metals	16.8
Fires involving electrical equipment	16.10
Fire involving gases	16.7
Fixed fire fighting systems	16.13
Flat roof as exit route	14.79
Floor	
— coverings	14.25, 20.2
— fire resistance of	12.1–12.2, Table A & Notes
— gradient of	14.32
— hatches	14.76
— openings in	12.7–12.8
Foliage, artificial and dried	20.4, Appendix B
Furniture and furnishings	20.3, 21.13 & Appendix B
Gate, wicket	14.80
General storage areas	13.13
Good housekeeping and the prevention of fire	Chapter 21
Goods delivery doors, shutters etc	14.81
Handrails	14.63, 14.71
Headroom of stairways, lobbies, corridors, passages	14.62
Health & Safety at Work etc Act 1974	11.2
Health and Safety Executive	11.2
Hearing impaired people	19.9–19.11
Heating appliances	14.7, 21.4–21.5
Helical stairways	14.64
High fire risk areas	13.8–13.11, 13.13, 14.18, 14.79(g), 14.80, 14.83, Tables A and D
Hoists	14.72
Hose reels	16.5
Improvement notices	4.10–4.14
Initial dead-end	14.17, diagram 4 and Table D
Inner and access rooms	14.18
Inspection of premises	3.6–3.9
Instruction and training	Chapter 17
Intercommunication between rooms and occupancies	14.77

Kitchens	13.11(d), 21.3(i), 21.4(c), 21.7–21.8
Ladder devices	14.67–14.68, 14.70
Lifts and hoists	Table A & Notes, 14.72–14.75 and diagram 21
List of fire precautions	21.4
Lighting	14.100–14.102
Lobbies	14.39, 14.73, diagrams 12 & 21
Local Acts relating to means of escape and fire precautions	Chapter 8
Location of fire fighting equipment	16.11–16.12
Lowering lines and other self rescue devices	14.69
Manipulative escape devices	14.69
Material alteration/extension	3.25–3.27
Means for giving warning in case of fire and for detecting fire	Chapter 15
Means for fighting fire	Chapter 16
Means of escape	Chapter 14
Mentally ill people	19.12
New buildings, structural alterations and the effect of the Building Regulations	Chapter 10
Obligations under fire certificates	3.17–3.18
Occupants of adjoining premises	14.77
Occupant capacity and occupant floor space factors	14.12, Table C & Notes
Offences	Chapter 6
Panic bolts and latches	14.88
Partitions, space dividers and other vertical surfaces	12.5
Passageways through rooms	14.25
People with disabilities	Chapter 19
Phased evacuation	15.6
Photo-luminescent signs or self luminous signs	Note to 14.94
Pictograms	14.94, 17.10
Plans of premises	3.5, 3.12
Portable heaters	14.7, 21.3
Premises	
— adjoining, means of escape from	14.77
— Crown	Chapter 9
— for which application has to be made for a fire certificate	Chapter 2
— in multiple occupation	1.7
Procedure for Fire Certification	Chapter 3
Prohibition notices	1.4, Chapters 5, 7.9
Protected lobby approach to single stairway building	14.39, diagram 12
Protected routes	Note to 14.3, Chapter 20
Public address system used for fire alarm	15.11 & Note
Ramps	14.71
Requests for further information	3.5
Responsibilities of management	Chapter 18
Revolving doors	14.90
Rights of appeal and grievances	Chapter 7
Roof exits and exits at upper levels	14.79
Routes of travel – angle of divergence	14.8, diagrams 1, 2 and Summary
Self-closing devices	14.84

Service duct work	12.7–12.8
Signs on doors	14.96–14.99
Sliding doors	14.92
Smoke spread	12.7–12.8, 13.4
Special fire risks associated with manufacturing processes	Introduction
Sprinklers	13.3 and Note, 13.14
Staff	
— alarms	15.7
— training	Chapters 17, 21
— with disabilities	Chapter 19
Staged fire alarms	15.4
Stairways	
— accommodation	14.56, 14.58
— by pass	14.43 & diagram 13
— enclosure of	14.42–14.50, diagrams 14, 15 & 16
— external	14.61
— number of	14.36–14.40
— protected routes	14.44, 14.46, Chapter 20, diagram 15
— spiral	14.64–14.66
— from basements	14.51–14.55 & diagram 18
— ventilation of	14.60
— width of	14.33–14.35, 14.41
Statutory interim duty pending disposal of application	3.3
Storage and display in shops	13.13
Surface finishes of walls, ceilings and escape routes	12.1–12.2, Table A, 12.3–12.5, Table B & examples
Surface spread of flame/classes	12.3–12.5, Table B & examples
Synthetic materials	Chapters 20, 21 & Appendix B
Tall items of plant	14.70
Toilets, fire resistance of door to	Table A & Notes
Training	Chapter 17
Travel distance	14.13–14.55, Table D & Notes
Travel from rooms to a storey exit	14.26–14.32
Travel within rooms	14.18–14.25
Travel within stairways and to final exits	14.33–14.55
Travolators	14.57–14.59
Two-stage evacuation	15.5
Unacceptable items within stairway enclosures	14.7
Unoccupied area	
— fire growth potential in	21.4, 21.14
Upholstered furniture	Appendix B
Upward escape	14.79
Ventilation	
— ducts	Table A & Notes, 12.7–12.8
— of stairways	14.60
— of lift well	14.75
— systems	14.103
— trunking	12.7–12.8
Vision impaired/blind people	19.7–19.8
Vision panels	Table A & Notes, 14.18, diagram 5, 14.19, 14.83(b)
Voids	12.7–12.8

Walkways	14.78
Wall	
— cavities	12.8
— fire resistance of	12.1–12.2, Table A & Notes
— hatches	14.76
Wheelchair users and people with impaired mobility	19.6
Wicket doors and gates	14.80
Width of exits from rooms, stairways etc	14.20–14.24, 14.41
Window exits	14.82, diagram 22